T0224603

Lecture Notes in Mathematics

Edited by J.-M. Morel, F. Takens and B. Teissier

Editorial Policy
for the publication of monographs

1. Lecture Notes aim to report new developments in all areas of mathematics – quickly, informally and at a high level. Monograph manuscripts should be reasonably self-contained and rounded off. Thus they may, and often will, present not only results of the author but also related work by other people. They may be based on specialized lecture courses. Furthermore, the manuscripts should provide sufficient motivation, examples and applications. This clearly distinguishes Lecture Notes from journal articles or technical reports which normally are very concise. Articles intended for a journal but too long to be accepted by most journals, usually do not have this "lecture notes" character. For similar reasons it is unusual for doctoral theses to be accepted for the Lecture Notes series.

2. Manuscripts should be submitted (preferably in duplicate) either to one of the series editors or to Springer-Verlag, Heidelberg. In general, manuscripts will be sent out to 2 external referees for evaluation. If a decision cannot yet be reached on the basis of the first 2 reports, further referees may be contacted: the author will be informed of this. A final decision to publish can be made only on the basis of the complete manuscript, however a refereeing process leading to a preliminary decision can be based on a pre-final or incomplete manuscript. The strict minimum amount of material that will be considered should include a detailed outline describing the planned contents of each chapter, a bibliography and several sample chapters.
Authors should be aware that incomplete or insufficiently close to final manuscripts almost always result in longer refereeing times and nevertheless unclear referees' recommendations, making further refereeing of a final draft necessary.
Authors should also be aware that parallel submission of their manuscript to another publisher while under consideration for LNM will in general lead to immediate rejection.

3. Manuscripts should in general be submitted in English.
Final manuscripts should contain at least 100 pages of mathematical text and should include
– a table of contents;
– an informative introduction, with adequate motivation and perhaps some
  historical remarks: it should be accessible to a reader not intimately familiar
  with the topic treated;
– a subject index: as a rule this is genuinely helpful for the reader.

Continued on inside back-cover

Lecture Notes in Mathematics 1785

Editors:
J.-M. Morel, Cachan
F. Takens, Groningen
B. Teissier, Paris

Springer

Berlin
Heidelberg
New York
Barcelona
Hong Kong
London
Milan
Paris
Tokyo

Juan Arias de Reyna

Pointwise Convergence
of Fourier Series

 Springer

Author

Juan Arias de Reyna
Facultad de Matemáticas
Universidad de Sevilla
Apdo. 1160
41080 Sevilla, Spain
e-mail: arias@us.es

Cataloging-in-Publication Data applied for

Die Deutsche Bibliothek - CIP-Einheitsaufnahme

Arias de Reyna, Juan:
Pointwise convergence of Fourier series / Juan Arias de Reyna. - Berlin ;
Heidelberg ; New York ; Barcelona ; Hong Kong ; London ; Milan ; Paris ;
Tokyo : Springer, 2002
  (Lecture notes in mathematics ; 1785)
  ISBN 3-540-43270-1

Mathematics Subject Classification (2000): 42A20

ISSN 0075-8434
ISBN 3-540-43270-1 Springer-Verlag Berlin Heidelberg New York

Springer-Verlag Berlin Heidelberg New York a member of BertelsmannSpringer
Science + Business Media GmbH

http://www.springer.de

© Springer-Verlag Berlin Heidelberg 2002
Printed in Germany

Typesetting: Camera-ready TeX output by the author

SPIN: 10868476   41/3142/DU - 543210 - Printed on acid-free paper

Preface

This book grew out of my attempt in August 1998 to compare Carleson's and Fefferman's proofs of the pointwise convergence of Fourier series with Lacey and Thiele's proof of the boundedness of the bilinear Hilbert transform. I started with Carleson's paper and soon realized that my summer vacation would not suffice to understand Carleson's proof.

Bit by bit I began to understand it. I was impressed by the breathtaking proof and started to give a detailed exposition that could be understandable by someone who, like me, was not a specialist in harmonic analysis. I've been working on this project for almost two years and lectured on it at the University of Seville from February to June 2000. Thus, this book is meant for graduate students who want to understand one of the great achievements of the twentieth century.

This is the first exposition of Carleson's theorem about the convergence of Fourier series in book form. It differs from the previous lecture notes, one by Mozzochi [38], and the other by Jørsboe and Mejlbro [26], in that our exposition points out the motivation of every step in the proof. Since its publication in 1966, the theorem has acquired a reputation of being an isolated result, very technical, and not profitable to study. There have also been many attempts to obtain the results by simpler methods. To this day it is the proof that gives the finest results about the maximal operator of Fourier series.

The Carleson analysis of the function, one of the fundamental steps of the proof, has an interesting musical interpretation. A sound wave consists of a periodic variation of pressure occurring around the equilibrium pressure prevailing at a particular time and place. The sound signal f is the variation of the pressure as a function of time. The Carleson analysis gives the score of a musical composition given the sound signal f. The Carleson analysis can be carried out at different levels. Obviously the above assertion is true only if we consider an adequate level.

Carleson's proof has something that reminds me of living organisms. The proof is based on many choices that seem arbitrary. This happens also in living organisms. An example is the **error** in the design of the eyes of the vertebrates. The photoreceptors are situated in the retina, but their outputs emerge on the **wrong** side: inside the eyes. Therefore the axons must finally be packed in the optic nerve that exit the eyes by the so called **blind spot**. But so many fibers (125 million light-sensitive cells) will not pass by a small spot. Hence evolution has solved the problem packing another layer of neurons inside the eyes that have rich interconections with the photoreceptors and with each other. These neurons process the information before it is send to the brain, hence the number of axons that must leave the eye is sustantially reduced (one million axons in each optic nerve). The incoming light must traverse these neurons to reach the photoreceptors, hence evolution has the added problem of making them transparent.

We have tried to arrange the proof so that these things do not happen, so that these arbitrary selections do not shade the idea of the proof. We have had the advantage of the text processor TeX, which has allowed us to rewrite without much pain. (We hope that no signs of these rewritings remain).

By the way, the eyes and the ears process the information in totally different ways. The proof of Carleson follows more the ear than the eyes. But what these neurons are doing in the inside of the eyes is just to solve the problem: How must I compress the information to send images using the least possible number of bits? A problem for which the wavelets are being used today.

I would like this book to be a commentary to the Carleson paper. Therefore we give the Carleson-Hunt theorem following more Carleson's than Hunt's paper.

The chapter on the maximal operator of Fourier series $S^* f$, gives the first exposition of the consequences of the Carleson-Hunt theorem. Some of the results appear here for the first time.

I wish to express my thanks to Fernando Soria and to N. Yu Antonov for sending me their papers and their comments about the consequences of the Carleson-Hunt theorem. Also to some members of the department of Mathematical Analysis of the University of Seville, especially to Luis Rodríguez-Piazza who showed me the example contained in chapter XIII.

Table of Contents

Introduction

The origin of Fourier series is the 18th century study by Euler and Daniel Bernoulli of the vibrating string. Bernoulli took the point of view, suggested by physical considerations, that every function can be expanded in a trigonometric series. At this time the prevalent idea was that such an expression implied differentiability properties, and that such an expansion was not possible in general.

Such a question was not one for that time. A response depended on what is understood by a function, a concept that was not clear until the 20th century. The first positive results were given in 1829 by Dirichlet, who proved that the expansion is valid for every continuous function with a finite number of maxima and minima.

A great portion of the mathematics of the first part of the 20th century was motivated by the convergence of Fourier series. For example, Cantor's set theory has its origin in the study of this convergence. Also Lebesgue's measure theory owes its success to its application to Fourier series.

Luzin, in 1913, while considering the properties of Hilbert's transform, conjectured that every function in $\mathcal{L}^2[-\pi, \pi]$ has an a. e. convergent Fourier series. Kolmogorov, in 1923 gave an example of a function in $\mathcal{L}^1[-\pi, \pi]$ with an a. e. divergent Fourier series.

A. P. Calderon (in 1959) proved that if the Fourier series of every function in $\mathcal{L}^2[-\pi, \pi]$ converges a. e., then

$$\mathfrak{m}\{x : \sup_n |S_n f(x)| > y\} \leq C \frac{\|f\|_2}{y^2}.$$

For many people the belief in Luzin's conjecture was destroyed; it seemed too good to be true.

So, it was a surprise when Carleson, in 1966, proved Luzin's conjecture. The next year Hunt proved the a. e. convergence of the Fourier series of every $f \in \mathcal{L}^p[-\pi, \pi]$ for $1 < p \leq \infty$.

Kolmogorov's example is in fact a function in $L \log \log L$ with a. e. divergent Fourier series. Hunt proved that every function in $L(\log L)^2$ has an a. e. convergent Fourier series. Sjölin, in 1969, sharpened this result: every function in the space $L \log L \log \log L$ has an a. e. convergent Fourier series. The last result in this direction is that of Antonov (in 1996) who proved the same for functions in $L \log L(\log \log \log L)$. Also, there are some quasi-Banach spaces of functions with a. e. convergent Fourier series given by Soria in 1985.

In the other direction, the results of Kolmogorov were sharpened by Chen in 1969 giving functions in $L(\log \log L)^{1-\varepsilon}$ with a. e. divergent Fourier series. Recently, Konyagin (1999) has obtained the same result for the space $L\varphi(L)$, whenever φ satisfies $\varphi(t) = o(\sqrt{(\log t / \log \log t)})$.

Apart from the proof of Carleson, there have been two others. First, the one by Fefferman in 1973. He says: *However, our proof is very inefficient near* L^1. *Carleson's construction can be pushed down as far as* $L \log L(\log \log L)$, *but our proof seems unavoidably restricted to* $L(\log L)^M$ *for some large* M. Then we have the recent proof of Lacey and Thiele; they are not so explicit as Fefferman, but what they prove (as far as I know) is limited to the case $p = 2$.

The proof of Lacey and Thiele is based on ideas from Fefferman proof, also the proof of Fefferman has been very important since it has inspired these two authors in his magnificent proof of the boundedness of the bilinear Hilbert transform. The trees and forests that appears in these proofs has some resemblance to the notes of the function and the allowed pairs of the proof of Carleson that are introduced in chapter eight and nine of this book, but to understand better these relationships will be matter for other book.

The aim of this book is the exposition of the principal result about the convergence of Fourier series, that is, the Carleson-Hunt Theorem.

The book has three parts. The first part gives a review of some results needed in the proof and consists of three chapters.

In the first chapter we give a review of the Hardy-Littlewod maximal function. We prove that this operator transforms $\mathcal{L}^p$ into $\mathcal{L}^p$ for $1 < p \le \infty$. The differentiation theorem allows one to see the great success we get with a pointwise convergence problem by applying the idea of the maximal function. This makes it reasonable to consider the maximal operator $S^* f(x) = \sup_n |S_n(f, x)|$, in the problem of convergence of Fourier series.

In chapter two we give elementary results about Fourier series. We see the relevance of the conjugate function and explain the elements on which Luzin, in 1913, founded his conjecture about the convergence of Fourier series of $\mathcal{L}^2$ functions. We also present Dini's and Jordan's tests of convergence in conformity with the law: s/he who does not know these criteria must not read Carleson's proof.

The properties of Hilbert's transform, needed in the proof of Carleson's Theorem are treated in chapter three.

In the second part we give the exposition of the Carleson-Hunt Theorem. The basic idea of the proof is the following. Our aim is to bound what we call Carleson integrals

$$\text{p.v.} \int_{-\pi}^{\pi} \frac{e^{in(x-t)}}{x - t} f(t) \, dt.$$

To this end we consider a partition Π of the interval $[-\pi, \pi]$ into subintervals, one of them $I(x)$ containing the point x, and write the integral as

$$\text{p.v.} \int_{-\pi}^{\pi} \frac{e^{in(x-t)}}{x-t} f(t)\, dt = \text{p.v.} \int_{I(x)} \frac{e^{in(x-t)}}{x-t} f(t)\, dt$$

$$+ \sum_{J \in \Pi, J \neq I(x)} \int_{J} \frac{e^{in(x-t)} f(t) - M_J}{x-t}\, dt + \int_{J} \frac{M_J}{x-t}\, dt.$$

where M_J is the mean value of $e^{in(x-t)} f(t)$ on the interval J.

The last sum can be conveniently bounded so that, in fact, we have changed the problem of bounding the first integral to the analogous problem for the integral on $I(x)$. After a change of scale we see that we have a similar integral, but the number of cycles in the exponent n has decreased in number.

Therefore we can repeat the reasoning. With this procedure we obtain the theorem that $S_n(f, x) = o(\log \log n)$ a. e., for every $f \in \mathcal{L}^2[-\pi, \pi]$. We think that to understand the proof of Carleson's theorem it is important to start with this theorem, because this is how the proof was generated. Only after we have understood this proof can we understand the very clever modifications that Carleson devised to obtain his theorem.

The next three chapters are dedicated to this end. The first deals with the actual bound of the second and third terms, and the problem of how we must choose the partition to optimize these bounds.

In Chapter five we prove that the bounds are good, with the exception of sets of controlled measure. Then, in Chapter six, given $f \in \mathcal{L}^2[-\pi, \pi]$, $y > 0$, $N \in N$, and $\varepsilon > 0$, we define a measurable set E with $\mathrm{m}(E) < A\varepsilon$ and such that

$$\sup_{0 \leq n \leq N/4} \left| \text{p.v.} \int_{-\pi}^{\pi} \frac{e^{in(x-t)}}{x-t} f(t)\, dt \right| \leq C \frac{\|f\|_2}{\sqrt{\varepsilon}} (\log N).$$

From this estimate we obtain the desired conclusion that

$$S_n(f, x) = o(\log \log n) \quad a. \ e.$$

These three chapters follow Carleson's paper, where instead of $f \in \mathcal{L}^2$ he assumed that $|f|(\log^+ |f|)^{1+\delta} \in \mathcal{L}^1$, reaching the same conclusion. Since we shall obtain further results, we have taken the simpler hypothesis that $f \in \mathcal{L}^2$. In fact, our motivation to include the proof is to allow the reader to understand the modifications contained in the next five chapters.

The logarithmic term appears in the above proof because every time we apply the basic procedure, we must put apart in the set E a small subset where the bound is not good. We have to put in a term $\log N$ in order to obtain a controlled measure. In fact, we are considering all pairs (n, J) formed by a dyadic subinterval J of $[-\pi, \pi]$, and the number of cycles of the Carleson integral. If we consider the procedure of chapters four to six, then we suspect that we do not need all of these pairs. This is the basic observation on which all the clever reasoning of Carleson is founded.

In chapter seven we determine which pairs are needed. Carleson made an analysis of the function to detect which pairs these are. If we think of f as the sound signal of a piece of music, then this analysis can be seen as a process to derive from f the score of this piece of music. In this chapter we define the set $\mathcal{Q}^j$ of notes of f to the level j.

In chapter eight we define the set $\mathcal{R}^j$ of allowed pairs. This is an enlargement of the set of notes of f, so that we can achieve two objectives. The principal objective is that if $\alpha = (n, J)$ is a pair such that $\alpha \notin \mathcal{R}^j$, then the sounds of the notes of f (at level j) that have a duration containing J, is essentially a single note or a rest. This is very important because if we consider a Carleson integral $\mathcal{C}_\alpha f(x)$ with this pair, then we have a candidate note, the sound of f, that is an allowed pair and therefore can be used in the basic procedure of chapter four.

Chapter nine is the most difficult part of the proof. In it we see how, given an arbitrary Carleson integral $\mathcal{C}_\alpha f(x)$, we can obtain an allowed pair ξ such that we can apply the procedure of chapter four, and a change of frequency to bound this integral.

In chapter ten we apply all this machinery to prove the basic inequality of Theorem 10.2.

The last part of the book is dedicated to deriving some consequences of the proof of Carleson-Hunt.

First, in chapter eleven we prove a version of the Marcinkiewicz interpolation theorem and give the definition and first properties of the spaces that we shall need in chapter twelve. In particular, we study a class of spaces near $\mathcal{L}^1(\mu)$ that play a prominent role. We prove that they are atomic spaces, a fact that allows very neat proofs in the following chapter.

In Chapter twelve we study the maximal operator S^*f of Fourier series. In it we give detailed and explicit versions of Hunt's theorem, with improved constants. We end the chapter by defining two quasi-Banach spaces, Q and QA, of functions with almost everywhere convergent Fourier series. These spaces improve the known results of Sjölin, Soria and Antonov, and the proofs are simpler.

In the last chapter we consider the Fourier transform on $\mathbf{R}$. We consider the problem of when we can obtain the Fourier transform of a function $f \in \mathcal{L}^p(\mathbf{R})$ by the formula

$$\widehat{f}(x) = \lim_{a \to +\infty} \int_{-a}^{a} f(t) e^{-2\pi i x t} \, dt.$$

We prove by an example (Example 13.2) that our results are optimal.

About the notation

I have tried to use standard notations. We use $\mathbf{N}$, $\mathbf{R}$, $\mathbf{C}$, $\mathcal{L}^p(\mathbf{R})$ in their usual meaning. $\mathbf{T}$ is the unit circle of the complex plane. $\mathfrak{m}$ denotes Lebesgue measure. The measure of a set A is sometimes written as $|A|$. When $B \subset \mathbf{R}^n$ is a ball of radius r, λB denotes the concentric ball of radius λr.

In chapter twelve we denote by μ normalized Lebesgue measure on $[-\pi, \pi]$. χ_A is the characteristic function of the set A.

Our notation is not standard when we denote

$$\|f\|_p = \left(\int_I |f|^p \, d\mathfrak{m} \right)^{1/p}, \qquad \|f\|_{\mathcal{L}^p(I)} = \left(\frac{1}{|I|} \int_I |f|^p \, d\mathfrak{m} \right)^{1/p},$$

where I denotes an interval. $\{0,1\}^*$ denotes the set of words in the alphabet $\{0,1\}$, therefore the set of finite (possibly empty) sequences of 0's and 1's.

At some points I have used Iverson-Knuth notation: $[P]$ is equal to 1 if P is true and to 0 if it is false. This is explained in p. 35.

A, B, C, ... a, b, c,... denote absolute constants. In the proof p usually denotes an exponent $1 < p < +\infty$. Every dependence on p is explicitly written (as in C_p). This has an exception: in the proof appears what we call the shift m. This is a natural number only depending on p. If we had put $m(p)$ or m_p it would have been very cumbersome.

Finally, we have tried to include all the notations in the following list.

List of Notations

Pairs

Maximal functions

Sets of pairs

Part One

Fourier Series and
Hilbert Transform

1. Hardy-Littlewood maximal function

1.1 Introduction

What Carleson proved in 1966 was Luzin's conjecture of 1913, and this proof depended on many results obtained in the fifty years since the conjecture was stated. In this chapter we make a rapid exposition of one of these prerequisites. We can also see one of the best ideas, that is, taking a maximal operator when one wants to prove pointwise convergence. The convergence result obtained is simple: the differentiability of the definite integral. This permits one to observe one of the pieces of Carleson's proof without any technical problems.

Given a function $f \in \mathcal{L}^1(\mathbf{R})$ we ask about the differentiability properties of the definite integral

$$F(x) = \int_{-\infty}^{x} f(t)\, dt.$$

This is equivalent to the question of whether there exists

$$\lim_{h \to 0} \frac{F(x+h) - F(x)}{h} = \lim_{h \to 0} \frac{1}{h} \int_{x}^{x+h} f(t)\, dt.$$

When we are confronted with questions of convergence it is advisable to study the corresponding maximal function. Here,

$$\sup_{h} \left| \frac{1}{h} \int_{x}^{x+h} f(t)\, dt \right|.$$

An analogous result in dimension n will be

$$f(x) = \lim_{Q \searrow x} \frac{1}{|Q|} \int_{Q} f(t)\, dt, \tag{1.1}$$

where Q denotes a cube of center x and side h and we write $Q \searrow x$ to express that the side $h \to 0^+$. In the one-dimensional case we have $Q = [x-h, x+h]$, this difference ($[x-h, x+h]$ instead of $[x, x+h]$) has no consequence, as we will see.

For every locally integrable function $f \colon \mathbf{R}^n \to \mathbf{C}$, we put

$$\mathcal{M}f(x) = \sup_Q \frac{1}{|Q|} \int_Q |f(t)|\, dt,$$

where the supremum is taken over all cubes $Q \subset \mathbf{R}^n$ with center x. $\mathcal{M}f$ is the Hardy-Littlewood maximal function.

1.2 Weak inequality

First observe that given f locally integrable, the function $\mathcal{M}f \colon \mathbf{R}^n \to [0, +\infty]$ is measurable. In fact for every positive real number α the set $\{\mathcal{M}f(x) > \alpha\}$ is open, because given $x \in \mathbf{R}^n$ with $\mathcal{M}f(x) > \alpha$ there exists a cube Q with center at x and such that

$$\frac{1}{|Q|} \int_Q |f(t)|\, dt > \alpha.$$

We only have to observe that the function

$$y \mapsto \frac{1}{|Q|} \int_{y+Q} |f(t)|\, dt$$

is continuous.

If $f \in \mathcal{L}^p(\mathbf{R}^n)$, with $1 < p < +\infty$, we shall show that $\mathcal{M}f \in \mathcal{L}^p(\mathbf{R}^n)$. However, for $p = 1$ this is no longer true. What we can say is only that f belongs to weak-$\mathcal{L}^1$. That is to say

$$\mathfrak{m}\{\mathcal{M}f(x) > \alpha\} \le c_n \frac{\|f\|_1}{\alpha}.$$

The proof is really wonderful. The set where $\mathcal{M}f(x) > \alpha$ is covered by cubes where the mean of $|f|$ is greater than α. If this set has a big measure, we shall have plenty of these cubes. Then we can select a big pairwise disjoint subfamily and this implies that the norm of f is big.

The most delicate point of this proof is that at which we select the disjoint cubes. This is accomplished by the following covering lemma

Lemma 1.1 (Covering lemma) *Let $\mathbf{R}^d$ be endowed with some norm, and let $c_d = 2 \cdot 3^d$. If $A \subset \mathbf{R}^d$ is a non-empty set of finite exterior measure, and $\mathcal{U}$ is a covering of A by open balls, then there is a finite subfamily of disjoint balls $B_1, \ldots, B_n$ of $\mathcal{U}$ such that*

$$c_d \sum_{j=1}^n \mathfrak{m}(B_j) \ge \mathfrak{m}^*(A).$$

Proof. We can assume that A is measurable, because if it were not, there would exist open set $G \supset A$ with $\mathfrak{m}(G)$ finite and such that $\mathcal{U}$ would be a covering of G. Now, assuming that A is measurable, there exists a compact set $K \subset A$ with $\mathrm{m}(K) \geq \mathrm{m}(A)/2$. Now select a finite subcovering of K, say that with the balls U_1, U_2, ..., U_m. Assume that these balls are ordered with decreasing radii. Then we select the balls B_j in the following way. First $B_1 = U_1$ is the greatest of them all. Then B_2 is the first ball in the sequence of U_j that is disjoint from B_1, if there is one, in the other case we put $n = 1$. Then B_3 will be the first ball from the U_j that is disjoint from $B_1 \cup B_2$. We continue in this way, until every ball from the sequence U_j has non-empty intersection with some B_j.

Now we claim that $K \subset \bigcup_{j=1}^{n} 3B_j$. In fact we know that $K \subset \bigcup_{j=1}^{m} U_j$. Hence for every $x \in K$, there is a first j such that $x \in U_j$. If this U_j is equal to some B_k obviously we have $x \in B_k \subset 3B_k$. In other case U_j intersects some $B_k = U_s$. Selecting the minimum k, it must be that $s < j$, for otherwise we would have selected U_j instead of B_k in our process. So the radius of the ball B_k is greater than or equal to that of U_j. It follows that $U_j \subset 3B_k$.

Therefore

$$\frac{1}{2}\mathrm{m}(A) \leq \mathrm{m}(K) \leq \sum_{j=1}^{n} 3^d \mathrm{m}(B_j),$$

and the construction implies that these balls are disjoint.   □

Lemma 1.2 (Hardy and Littlewood) *If $f \in \mathcal{L}^1(\mathbf{R}^d)$ then $\mathcal{M}f$ satisfies, for each $\alpha > 0$, the weak inequality*

$$\mathrm{m}\{x \in \mathbf{R}^d \mid \mathcal{M}f(x) > \alpha\} \leq c_d \frac{\|f\|_1}{\alpha}.$$

Proof. Let $A = \{x \in \mathbf{R}^d \mid \mathcal{M}f(x) > \alpha\}$, it is an open set. We do not know yet that it has finite measure, so we consider $A_n = A \cap B_n$, where B_n is a ball of radius n and center 0. Now each $x \in A_n$ has $\mathcal{M}f(x) > \alpha$; hence there exists an open cube Q, with center at x and such that

$$\frac{1}{|Q|} \int_Q |f(t)| \, dt > \alpha. \tag{1.2}$$

Now cubes are balls for the norm $\| \cdot \|_\infty$ on $\mathbf{R}^d$. So we can apply the covering lemma to obtain a finite set of **disjoint** cubes $(Q_j)_{j=1}^{m}$ such that every one of them satisfies (1.2), and

$$\mathrm{m}(A_n) \leq c_d \sum_{j=1}^{m} \mathrm{m}(Q_j).$$

Therefore we have

$$\mathfrak{m}(A_n) \leq c_d \sum_{j=1}^{m} \frac{1}{\alpha} \int_Q |f(t)| \, dt.$$

Since the cubes are disjoint

$$\mathfrak{m}(A_n) \leq c_d \frac{\|f\|_1}{\alpha}.$$

Taking limits when $n \to \infty$, we obtain our desired bound.   □

1.3 Differentiability

As an application we desire to obtain (1.1). In fact we can prove something more. It is not only that at almost every point $x \in \mathbf{R}^d$ we have

$$\lim_{Q \searrow x} \frac{1}{|Q|} \left| \int_Q (f(t) - f(x)) \, dt \right| = 0,$$

but that we have

$$\lim_{Q \searrow x} \frac{1}{|Q|} \int_Q |f(t) - f(x)| \, dt = 0.$$

A point where this is true is called a **Lebesgue point** of f.

Theorem 1.3 (Differentiability Theorem) *Let $f \colon \mathbf{R}^d \to \mathbf{C}$ be a locally integrable function. There exists a subset $Z \subset \mathbf{R}^d$ of null measure and such that every $x \notin Z$ is a Lebesgue point of f. That is*

$$\lim_{Q \searrow x} \frac{1}{|Q|} \int_Q |f(t) - f(x)| \, dt = 0.$$

Proof. Whether x is a Lebesgue point of f or not, depends only on the values of f in a neighborhood of x. So we can reduce to the case of f integrable.

Also the results are true for a dense set on $\mathcal{L}^1(\mathbf{R}^d)$. In fact if f is continuous, given x and $\varepsilon > 0$, there is a neighborhood of x such that $|f(t) - f(x)| < \varepsilon$. Hence if Q denotes a cube with a sufficiently small radius we have

$$\frac{1}{|Q|} \int_Q |f(t) - f(x)| \, dt \leq \varepsilon.$$

Hence, for a continuous function f, every point is a Lebesgue point.

Now we can observe for the first time how the maximal function intervenes in pointwise convergence matters.

We are going to define the operator Ω. If $f \in \mathcal{L}^1(\mathbf{R}^d)$,

$$\Omega f(x) = \limsup_{Q \searrow x} \frac{1}{|Q|} \int_Q |f(t) - f(x)| \, dt$$

Note that
$$\Omega f(x) \le \mathcal{M}f(x) + |f(x)|.$$

Now our objective is to prove that $\Omega f(x) = 0$ almost everywhere.

Fix $\varepsilon > 0$. Since the continuous functions are dense on $\mathcal{L}^1(\mathbf{R}^d)$, we obtain a continuous $\varphi \in \mathcal{L}^1(\mathbf{R}^d)$, such that $\|f - \varphi\|_1 < \varepsilon$.

By the triangle inequality
$$\begin{aligned}
\Omega f(x) &\le \Omega\varphi(x) + \Omega(f - \varphi)(x) = \Omega(f - \varphi)(x) \\
&\le \mathcal{M}(f - \varphi)(x) + |f(x) - \varphi(x)|.
\end{aligned}$$

Hence for every $\alpha > 0$ we have
$$\{\Omega f(x) > \alpha\} \subset \{\mathcal{M}(f - \varphi)(x) > \alpha/2\} \cup \{|f(x) - \varphi(x)| > \alpha/2\}.$$

Now we use the weak inequality for the Hardy-Littlewood maximal function and the Chebyshev inequality for $|f - \varphi|$
$$\mathfrak{m}\{\Omega f(x) > \alpha\} \le 2c_d \frac{\|f - \varphi\|_1}{\alpha} + 2\frac{\|f - \varphi\|_1}{\alpha} \le C_d \frac{\varepsilon}{\alpha}.$$

Since this inequality is true for every $\varepsilon > 0$, we deduce $\mathfrak{m}\{\Omega f(x) > \alpha\} = 0$. And this is true for every $\alpha > 0$, hence $\Omega f(x) = 0$ almost everywhere. $\square$

As an example we prove that
$$F(x) = \int_{-\infty}^{x} f(t)\, dt$$

is differentiable at every Lebesgue point of f. We assume that f is integrable.

For $h > 0$
$$\frac{F(x+h) - F(x)}{h} - f(x) = \frac{1}{h}\int_x^{x+h} \left(f(t) - f(x)\right) dt.$$

Hence
$$\begin{aligned}
\left|\frac{F(x+h) - F(x)}{h} - f(x)\right| &\le \frac{1}{h}\int_x^{x+h} \left|f(t) - f(x)\right| dt \\
&\le \frac{2}{2h}\int_{x-h}^{x+h} \left|f(t) - f(x)\right| dt.
\end{aligned}$$

If x is a Lebesgue point of f we know that the limit when $h \to 0$ is equal to zero. An analogous procedure proves the existence of the left-hand limit at x.

1.4 Interpolation

At one extreme, with $p = 1$, the maximal function $\mathcal{M}f$ satisfies a weak inequality. At the other extreme $p = +\infty$, it is obvious from the definition that if $f \in \mathcal{L}^\infty(\mathbf{R}^d)$

$$\|\mathcal{M}f\|_\infty \le \|f\|_\infty.$$

An idea of Marcinkiewicz permits us to interpolate between these two extremes.

Theorem 1.4 *For every $f \in \mathcal{L}^p(\mathbf{R}^d)$, $1 < p < +\infty$ we have*

$$\|\mathcal{M}f\|_p \le C_d \frac{p}{p-1} \|f\|_p.$$

Proof. For every $\alpha > 0$ we decompose f, $f = f\chi_A + f\chi_{\mathbf{R}^d \setminus A}$, where $A = \{|f| > \alpha\}$. Then $\mathcal{M}f \le \alpha + \mathcal{M}(f\chi_A)$. Consequently

$$\mathfrak{m}\{\mathcal{M}f > 2\alpha\} \le \mathfrak{m}\{\mathcal{M}(f\chi_A) > \alpha\} \le \frac{c_d}{\alpha} \int_{\mathbf{R}^d} |f| \chi_{\{|f|>\alpha\}} \, d\mathfrak{m}.$$

The proof depends on a judicious use of this inequality. In particular observe that we have used a different decomposition of f for every α.

We have the following chain of inequalities

$$\|\mathcal{M}f\|_p^p = p \int_0^{+\infty} t^{p-1} \mathfrak{m}\{\mathcal{M}f > t\} \, dt \le$$

$$p \int_0^{+\infty} t^{p-1} \frac{2c_d}{t} \int_{\mathbf{R}^d} |f| \chi_{\{|f|>t/2\}} \, d\mathfrak{m} \, dt \le$$

Applying Fubini's theorem

$$2c_d p \int_{\mathbf{R}^d} |f(x)| \int_0^{+\infty} t^{p-2} \chi_{\{|f(x)|>t/2\}} \, dt \, dx =$$

$$2c_d p \int_{\mathbf{R}^d} \frac{(2|f(x)|)^{p-1}}{p-1} |f(x)| \, dx = \frac{2^p c_d p}{p-1} \|f\|_p^p$$

It is easy to see that $(p/(p-1))^{1/p}$ is equivalent to $p/(p-1)$. Hence we obtain our claim about the norm.   □

In the case of $p = 1$ the best we can say is the weak inequality. For example if $\|f\|_1 > 0$, then $\mathcal{M}f$ is not integrable. In spite of this we shall need in the proof of Carleson theorem a bound of the integral of the maximal function on a set of finite measure; that is a consequence of the weak inequality

Proposition 1.5 *For every function $f \in \mathcal{L}^1(\mathbf{R}^d)$ and $B \subset \mathbf{R}^d$ a measurable set*

$$\int_B \mathcal{M}f(x)\,dx \leq \mathbf{m}(B) + 2c_d \int_{\mathbf{R}^d} |f(x)|\log^+|f(x)|\,dx.$$

Proof. Let $\mathbf{m}_B$ be the measure $\mathbf{m}_B(M) = \mathbf{m}(B \cap M)$. We have

$$\int_B \mathcal{M}f(x)\,dx = \int_0^{+\infty} \mathbf{m}_B\{\mathcal{M}f(x) > t\}\,dt.$$

Now we have two inequalities $\mathbf{m}_B\{\mathcal{M}f(x) > t\} \leq \mathbf{m}(B)$, and the weak inequality.

The point of the proof is to use adequately the weak inequality. For every α we have $f = f\chi_A + f\chi_{\mathbf{R}^d \smallsetminus A}$ where $A = \{f(x) > \alpha\}$. Therefore $\mathcal{M}f \leq \alpha + \mathcal{M}(f\chi_A)$, and

$$\{\mathcal{M}f(x) > 2\alpha\} \subset \{\mathcal{M}(f\chi_A)(x) > \alpha\}.$$

It follows that

$$\mathbf{m}\{\mathcal{M}f(x) > 2\alpha\} \leq \frac{c_d}{\alpha} \int_{\{|f(x)|>\alpha\}} |f(x)|\,dx.$$

Hence

$$\int_B \mathcal{M}f(x)\,dx \leq \mathbf{m}(B) + 2\int_1^{+\infty} \frac{c_d}{t}\left(\int_{\{|f(x)|>t\}} |f(x)|\,dx\right)dt.$$

Therefore by Fubini's theorem

$$\int_B \mathcal{M}f(x)\,dx \leq \mathbf{m}(B) + 2c_d \int_{\mathbf{R}^d} |f(x)|\log^+|f(x)|\,dx.$$

$\square$

1.5 A general inequality

The Hardy-Littlewood maximal function can be used to prove many theorems of pointwise convergence. This and many other applications of these functions derive from the following inequality.

Theorem 1.6 *Let $\varphi\colon \mathbf{R}^d \to \mathbf{R}$ be a positive, radial, decreasing, and integrable function. Then for every $f \in \mathcal{L}^p(\mathbf{R}^d)$ and $x \in \mathbf{R}^d$ we have*

$$|\varphi * f(x)| \leq C_d \|\varphi\|_1 \mathcal{M}f(x),$$

where C_d is a constant depending only on the dimension, and equal to 1 for $d = 1$.

Proof. We say that φ is radial if there is a function $u \colon [0, +\infty) \to \mathbf{R}$ such that $\varphi(x) = u(|x|)$ for every $x \in \mathbf{R}^d$. Also we say that a radial function φ is decreasing if u is decreasing.

The function u is measurable, hence there is an increasing sequence of simple functions (u_n) such that $u_n(t)$ converges to $u(t)$ for every $t \geq 0$. In this case, since u is decreasing, it is possible to choose each u_n

$$u_n(t) = \sum_{j=1}^{N} h_j \, \chi_{[0,t_j]}(t),$$

where $0 < t_1 < t_2 < \cdots < t_N$ and $h_j > 0$ and the natural number N depends on n.

Now the proof is straightforward. Let $\varphi_n(x) = u_n(|x|)$. By the monotone convergence theorem

$$|\varphi * f(x)| \leq \varphi * |f|(x) = \lim_n \varphi_n * |f|(x).$$

Therefore

$$\varphi_n * |f|(x) = \sum_{j=1}^{N} h_j \int_{B(x,t_j)} |f(y)| \, dy.$$

We can replace the ball $B(x, t_j)$ by the cube with center x and side $2t_j$. The quotient between the volume of the ball and the cube is bounded by a constant. Thus

$$\varphi_n * |f|(x) \leq \sum_{j=1}^{N} h_j \mathfrak{m}\big(Q(x_j, t_j)\big) \cdot \mathcal{M}f(x) \leq C_d \|\varphi\|_1 \mathcal{M}f(x).$$

$\square$

2. Fourier Series

2.1 Introduction

Let $f: \mathbf{R} \to \mathbf{C}$ be a 2π-periodic function, integrable in $[-\pi, \pi]$. The Fourier series of f is the series

$$\sum_{j=-\infty}^{+\infty} a_j e^{ijt} \tag{2.1}$$

where the Fourier coefficients a_j are defined by

$$a_j = \frac{1}{2\pi} \int_{-\pi}^{\pi} f(t) e^{-ijt}\, dt. \tag{2.2}$$

These coefficients are denoted as $\widehat{f}(j) = a_j$.

These series had been considered in the eighteen century by Daniel Bernoulli, Euler, Lagrange, etc. They knew that if a function is given by the series (2.1), the coefficients can be calculated by (2.2). They also knew many examples.

Bernoulli, studying the movement of a string fixed at its extremes, gave the expression

$$y(x, t) = \sum_{j=1}^{\infty} a_j \sin \frac{j\pi x}{\ell} \cos j\rho t,$$

for its position, where ℓ is the length of the string and the coefficient ρ depends on its physical properties.

In 1753 Euler noticed a paradoxical implication: the initial position of the string would be given by

$$f(x) = \sum_{j=1}^{\infty} a_j \sin \frac{j\pi x}{\ell}.$$

At this moment the curves were classified as *continuous*, if they were defined by a formula, and *geometrical* if they could be drawn with the hand. They thought that the first ones were locally determined while the movement of the hand was not determined by the first stroke. Bernoulli believed that the representation of an arbitrary function was possible.

Fourier affirmed in his book *Théorie Analytique de la Chaleur* (1822) that the development was valid in the general case. This topic is connected with the definition of the concept of function.

2.2 Dirichlet Kernel

The convergence of the series (2.1) was considered by Dirichlet in 1829. He proved that the series converges to $(f(x+0)+f(x-0))/2$ for every piecewise continuous and monotonous function. This was later superseded by the results of Dini and Jordan. To prove these results we consider first the result of Riemann

Proposition 2.1 (Riemann-Lebesgue lemma) *If $f: \mathbf{R} \to \mathbf{C}$ is 2π-periodic and integrable on $[-\pi, \pi]$, then*

$$\lim_{|j| \to \infty} \widehat{f}(j) = 0.$$

Proof. If we change variables $u = t + \pi/j$ in the integral (2.2) the exponential changes sign. Hence we have

$$\widehat{f}(j) = \frac{1}{4\pi} \int_{-\pi}^{\pi} f(t) e^{-ijt}\, dt - \frac{1}{4\pi} \int_{-\pi}^{\pi} f\left(t - \frac{\pi}{j}\right) e^{-ijt}\, dt.$$

For a continuous function f it follows that $\lim |\widehat{f}(j)| = 0$. For a general f we approximate it in $\mathcal{L}^1$ norm by a continuous function.   □

To study pointwise convergence we consider the partial sums

$$S_n(f, x) = \sum_{j=-n}^{n} \hat{f}(j) e^{ijx}.$$

Since every coefficient has an integral expression, we obtain an integral form for the partial sum of the Fourier series

$$S_n(f, x) = \frac{1}{2\pi} \int_{-\pi}^{\pi} D_n(x - t) f(t)\, dt,$$

where the function D_n, the **Dirichlet kernel**, is given by

$$D_n(t) = \sum_{j=-n}^{n} e^{ijt} = \frac{\sin\left(n + \frac{1}{2}\right) t}{\sin t/2}.$$

It follows that $f \mapsto S_n(f, x)$ is a continuous linear form defined on $\mathcal{L}^1[-\pi, \pi]$.

The function D_n is 2π-periodic, with integral equal to 1, but $\|D_n\|_1$ and $\|D_n\|_\infty$ are not uniformly bounded.

With the integral expression of the partial sums we can obtain the two basic conditions for pointwise convergence.

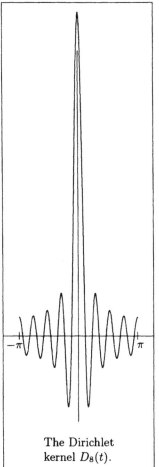

The Dirichlet kernel $D_8(t)$.

Theorem 2.2 (Dini's test) *If* $f \in \mathcal{L}^1[-\pi, \pi]$ *and*

$$\int_0^{\pi} |f(x + t) + f(x - t) - 2f(x)| \frac{dt}{t} < +\infty,$$

then the Fourier series of f *at the point* x *converges to* $f(x)$.

Proof. The difference $S_n(f) - f$ can be written as

$$S_n(f, x) - f(x) = \frac{1}{2\pi} \int_{-\pi}^{\pi} D_n(t)\big(f(x - t) - f(x)\big)\, dt =$$

$$\frac{1}{2\pi} \int_0^{\pi} D_n(t)\big(f(x + t) + f(x - t) - 2f(x)\big)\, dt.$$

Since $2 \sin t/2 \sim t$, the Riemann-Lebesgue lemma proves that this difference tends to 0. □

Theorem 2.3 (Jordan's test) *If $f \in \mathcal{L}^1[-\pi, \pi]$ is of bounded variation on an open interval that contains x, then the Fourier series at x converges to $\big(f(x+0) + f(x-0)\big)/2$.*

Proof. The proof is based on the fact that although $\|D_n\|_1$ are not bounded the integrals

$$\int_0^\delta D_n(t)\, dt$$

are uniformly bounded on n and δ. (This can be proved changing the Dirichlet kernel to the equivalent $\sin\left(n + \frac{1}{2}\right)/t$ and then applying a change of variables).

Without loss of generality we can assume that $x = 0$. Also we can assume that f is increasing on a neighborhood of 0. We must prove

$$\lim_n \frac{1}{2\pi} \int_0^\pi D_n(t)\big(f(t) + f(-t)\big)\, dt = \big(f(0+) + f(0-)\big)/2.$$

By symmetry it suffices to prove

$$\lim_n \frac{1}{2\pi} \int_0^\pi D_n(t) f(t)\, dt = f(0+)/2.$$

Finally we can assume that $f(0+) = 0$.

Choose $\delta > 0$ such that $0 \le f(t) < \varepsilon$ for every $0 < t < \delta$. We decompose the integral into two parts, one on $[0, \delta]$ and the other on $[\delta, \pi]$. We apply to the first integral the second mean value theorem, which states that if g is continuous and f monotone on $[a, b]$, there exists $c \in [a, b]$ such that

$$\int_a^b f(t) g(t)\, dt = f(b-) \int_c^b g(t)\, dt + f(a+) \int_a^c g(t)\, dt.$$

Therefore

$$\int_0^\pi D_n(t) f(t)\, dt = f(\delta-) \int_\eta^\delta D_n(t)\, dt + \int_\delta^\pi D_n(t) f(t)\, dt.$$

The second integral converges to 0 by the Riemann-Lebesgue lemma and the first is less than $C\varepsilon$ by the property of Dirichlet kernel that we have noted. $\qquad \square$

We see that these conditions only depend on the values of f in an arbitrarily small neighborhood of x. This is a general fact and it is known as the Riemann localization principle: the convergence of the Fourier series to $f(x)$ only depends on the values of f in a neighborhood of f. This is clear from the expression of $S_n(f, x)$ as an integral and the Riemann-Lebesgue lemma. This is surprising, because each $\widehat{f}(j)$ depends on all the values of f.

The two criteria given are independent. If $f(t) = 1/|\log(t/2\pi)|$, $g(t) = t^\alpha \sin(1/t)$, for $0 < t < \pi$ and $0 < \alpha < 1$, then f satisfies Jordan's condition but not Dini's test, at the point $t = 0$. On the other hand g satisfies only Dini's test.

2.3 Fourier series of continuous functions

The convergence conditions that we have proved show that the Fourier series of a differentiable function converges pointwise to the function. This is not true for continuous functions. Du Bois Reymond constructed a continuous function whose Fourier series is divergent at one point.

This follows from Banach-Steinhauss theorem. We consider $T_n(f) = S_n(f, 0)$ as a linear operator on the space of continuous functions on $[-\pi, \pi]$ that take the same values at the extremes. By the Banach-Steinhauss theorem $\sup_n \|T_n\| < +\infty$ if and only if for every $f \in \mathcal{C}(\mathbf{T})$ we have $\sup_N |T_n(f)| < +\infty$. But an easy calculus shows that

$$\|T_n\| = \|D_n\|_1 = \frac{4}{\pi^2} \log n + O(1).$$

The numbers $L_n = \|D_n\|_1$ are called Lebesgue constants. Its order is computed as follows

$$L_n = \frac{2}{2\pi} \int_0^\pi \left| \frac{\sin(n + 1/2)t}{\sin(t/2)} \right| dt = \frac{2}{\pi} \int_0^\pi \left| \frac{\sin(n + 1/2)t}{t} \right| dt + O(1)$$

$$= \frac{2}{\pi} \int_0^{(2n+1)\pi/2} \left| \frac{\sin u}{u} \right| du + O(1) = \frac{2}{\pi} \sum_{k=0}^{n-1} \int_{k\pi}^{(k+1)\pi} \left| \frac{\sin u}{u} \right| du + O(1)$$

$$= \frac{2}{\pi} \sum_{k=0}^{n-1} \int_0^\pi \left| \frac{\sin u}{k\pi + u} \right| du + O(1)$$

$$\frac{2}{\pi} \int_0^\pi \sum_{k=1}^{n-1} \left| \frac{\sin u}{k\pi + u} \right| du + O(1) = \frac{4}{\pi} \sum_{k=1}^{n-1} \frac{1}{k\pi + \xi} + O(1) = \frac{4}{\pi^2} \log n + O(1).$$

Notice the following corollary

Corollary 2.4 If $f \in \mathcal{L}^\infty[-\pi, \pi]$, then $|S_n(f, x)| \leq (\frac{4}{\pi^2} \log n + C)\|f\|_\infty$.

The following theorem is more difficult. In its proof we need an expression of the Dirichlet kernel that plays an important role in Carleson's theorem.

Theorem 2.5 (Hardy) *If $f \in \mathcal{L}^1[-\pi, \pi]$, then at every Lebesgue point x of f*

$$\lim_{n \to \infty} [S_n(f, x)/(\log n)] = 0.$$

Furthermore, if f is continuous on an open interval I, the convergence is uniform on every closed $J \subset I$.

Proof. The Dirichlet kernel can be written as

$$D_n(t) = \frac{\sin\left(n + \frac{1}{2}\right)t}{\sin t/2} = 2\frac{\sin nt}{t} + \cos nt + \left(\frac{1}{\tan t/2} - \frac{2}{t}\right)\sin nt.$$

The last two terms are bounded uniformly in n and t. Therefore

$$D_n(t) = 2\frac{\sin nt}{t} + \varphi_n(t), \qquad |t| < \pi, \tag{2.3}$$

and there is an absolute constant $0 < C < +\infty$ such that $\|\varphi_n\|_\infty \le C$. This expression will play a role in Carleson's theorem.

Now we have

$$\left| S_n(f, x) - \frac{1}{\pi}\int_{-\pi}^{\pi} f(x - t)\frac{\sin nt}{t}\, dt \right| \le c \int_{-\pi}^{\pi} |f(t)|\, dt.$$

It follows that for every f in $\mathcal{L}^1[-\pi, \pi]$

$$|S_n(f, x)| \le c\|f\|_1 + \frac{1}{\pi}\left| \int_{-\pi}^{\pi} f(x - t)\frac{\sin nt}{t}\, dt \right|.$$

The function $\sin t/t$ has integrals uniformly bounded on intervals. It follows that

$$|S_n(f, x)| \le C + \frac{1}{\pi}\left| \int_0^{\pi} \{f(x + t) + f(x - t) - 2f(x)\}\frac{\sin nt}{t}\, dt \right|.$$

Let $\varphi_x(t)$ denote the function $f(x+t) + f(x-t) - 2f(x)$. If x is a Lebesgue point of f, the primitive $\Phi(t)$ of $|\varphi_x(t)|$ satisfies $\Phi(t) = o(t)$ when $t \to 0$. With these notations

$$|S_n(f, x)| \le C + \frac{n}{\pi}\int_0^{1/n} |\varphi_x(t)|\, dt + \frac{1}{\pi}\int_{1/n}^{\pi} t^{-1}|\varphi_x(t)|\, dt$$

$$= C + \frac{n}{\pi}\Phi\left(\frac{1}{n}\right) + \frac{1}{\pi}\left.\Phi(t)t^{-1}\right|_{1/n}^{\pi} + \frac{1}{\pi}\int_{1/n}^{\pi} \Phi(t)t^{-2}\, dt.$$

It follows easily that

$$S_n(f, x)/\log n \to 0$$

since, x being a Lebesgue point of f, we have $\Phi(t) = o(t)$. $\qquad\square$

This result is the best possible in the following sense: for every sequence (λ_n) such that $\lambda_n^{-1} \log n \to +\infty$, there exists a continuous function f such that $|S_n(f, x)| > \lambda_n$ for infinitely many natural numbers n.

Some properties of the trigonometrical system follow from the fact that it is an orthonormal set of functions. For example D. E. Menshov and H. Rademacher proved that the series of orthonormal functions $\sum_j c_j \varphi_j$ converges almost everywhere if $\sum_j |c_j \log j|^2 < +\infty$. This is also the best result for general orthonormal systems, in particular D. E. Menshov in 1923 proved that there exist series $\sum_j c_j \varphi_j$ divergent almost everywhere and such that $\sum_j |c_j|^2 < +\infty$. Therefore Carleson's theorem is a property of the trigonometrical system that depends on the natural ordering of this system.

The result of D. E. Menshov and H. Rademacher is easily proved. In fact this is a general result that relates the order of $S_n(f, x)$ and the convergence of certain series.

Theorem 2.6 *Assume that (λ_n) is an increasing sequence of positive real number such that for every $g \in \mathcal{L}^2[-\pi, \pi]$*

$$\lim_{n \to +\infty} \frac{S_n(g, x)}{\lambda_{n+1}} = 0.$$

Then if $f \in \mathcal{L}^1[-\pi, \pi]$ satisfies $\sum_j |\widehat{f}(j) \lambda_{|j|}|^2 < +\infty$, then

$$f(x) = \lim_n S_n(f, x), \qquad a.\ e.$$

Proof. By the Riesz-Fischer theorem there exists a function $g \in \mathcal{L}^2[-\pi, \pi]$, such that $\widehat{g}(j) = \widehat{f}(j) \lambda_{|j|}$. Comparing Fourier coefficients we derive the equality

$$S_n(f, x) = \sum_{k=0}^{n} \left(\frac{1}{\lambda_k} - \frac{1}{\lambda_{k+1}} \right) S_k(g, x) + \frac{S_n(g, x)}{\lambda_{n+1}}.$$

By our hypothesis about the functions on $\mathcal{L}^2[-\pi, \pi]$, we deduce that the character of the sequence $S_n(f, x)$ coincides with that of the series

$$\sum_{k=1}^{\infty} \left(\frac{1}{\lambda_k} - \frac{1}{\lambda_{k+1}} \right) S_k(g, x) = \sum_{k=1}^{\infty} h_k(x).$$

But as a series on $\mathcal{L}^2[-\pi, \pi]$, we have $\sum_k \|h_k\|_2 < +\infty$. Therefore the series converges a. e. □

2.4 Banach continuity principle

The hypothesis that $\lim_n S_n(f,x)/\lambda_{n+1} \to 0$ in the theorem 2.6 can be replaced by $\sup_n |S_n(f,x)/\lambda_{n+1}| < +\infty$ a. e. This is a general fact due to Banach. This reduces the problem of a. e. convergence of Fourier series to proving the pointwise boundedness of the maximal operator

$$\sup_n |S_n(f,x)|.$$

To prove these assertions we need some knowledge about the space of measurable functions $\mathcal{L}^0[-\pi,\pi]$. It is a metric space with distance

$$d(f,g) = \frac{1}{2\pi} \int_{-\pi}^{\pi} \frac{|f-g|}{1+|f-g|} \, dm.$$

This is a complete metric vector space. A sequence (f_n) converges to 0 if and only if it converges to 0 in measure. That is to say: for every $\varepsilon > 0$ we have $\lim_n m\{|f_n| > \varepsilon\} = 0$.

Consider now a sequence (T_n) of linear operators

$$T_n \colon \mathcal{L}^p[-\pi,\pi] \to \mathcal{L}^0[-\pi,\pi].$$

We assume that each T_n is continuous in measure, therefore for every (f_k) with $\|f_k\|_p \to 0$, and every $\varepsilon > 0$ we have $m\{|T(f_k)| > \varepsilon\} \to 0$, (and this is true for every $T = T_n$). Observe that if $T_n f(x)$ converges a. e., then the maximal operator $T^* f(x) = \sup_n |T_n f(x)|$ is bounded a. e.

The principle of Banach is a sort of uniform boundedness principle: The continuity in measure of a sequence of operators and the almost everywhere finiteness of the maximal operator imply the continuity at 0 in measure of the maximal operator.

Theorem 2.7 (Banach's continuity principle) *Let us assume that for every $f \in \mathcal{L}^p[-\pi,\pi]$, the function $T^* f(x) < +\infty$ a. e. on $[-\pi,\pi]$, then there exists a decreasing function $C(\alpha)$ defined for every $\alpha > 0$, such that $\lim_{\alpha \to +\infty} C(\alpha) = 0$ and such that*

$$m\{T^* f(x) > \alpha \|f\|_p\} \le C(\alpha),$$

for every $f \in \mathcal{L}^p[-\pi,\pi]$

Proof. Fix a positive real number $\varepsilon > 0$. For every natural number let F_n be the set of $f \in \mathcal{L}^p[-\pi,\pi]$ such that $m\{T^* f(x) > n\} \le \varepsilon$.

The set F_n is closed on $\mathcal{L}^p[-\pi,\pi]$. To prove this consider $f \notin F_n$, then

$$m\{T^* f(x) > n\} > \varepsilon.$$

It follows that there exists N such that

$$\mathfrak{m}\{ \sup_{1 \le k \le N} T_k f(x) > n \} > \varepsilon.$$

Then there exists $\delta > 0$ such that

$$\mathfrak{m}\{ \sup_{1 \le k \le N} T_k f(x) > n + \delta \} > \varepsilon + \delta.$$

By the continuity in measure of the operators T_k, there exists a $\delta' > 0$ such that for every g with $\|f - g\|_p < \delta'$ we have

$$\mathfrak{m}\{ |T_k(f - g)(x)| > \delta \} < \delta/2^k, \qquad 1 \le k \le N.$$

Let Z be the union of the exceptional sets $\{ |T_k(f-g)(x)| > \delta \}$. Then $\mathfrak{m}(Z) < \delta$. Also we have

$$\{ T^* g(x) > n \} \cup Z \supset \{ \sup_{1 \le k \le N} T_k f(x) > n + \delta \}.$$

Therefore it follows that

$$\mathfrak{m}\{ T^* g(x) > n \} > \varepsilon.$$

That is, the set $\mathcal{L}^p[-\pi, \pi] \smallsetminus F_n$ is open.

Now our hypothesis about the boundedness of $T^* f$ implies that

$$\mathcal{L}^p[-\pi, \pi] = \bigcup_n F_n.$$

By Baire's category theorem there is some $n \in \mathbf{N}$ such that F_n has a non-empty interior. That is, there exist $f_0 \in F_n$ and $\delta > 0$, such that $f = f_0 + \delta g$ with $\|g\|_p = 1$. Thus

$$\mathfrak{m}\{ T^*(f_0 + \delta g) > n \} \le \varepsilon.$$

Then

$$\mathfrak{m}\{ T^* g > 2n/\delta \} \le \mathfrak{m}\{ T^*(f_0 + \delta g) > n \} + \mathfrak{m}\{ T^*(f_0 - \delta g) > n \} \le 2\varepsilon.$$

Therefore for every $g \in \mathcal{L}^p[-\pi, \pi]$

$$\mathfrak{m}\{ T^* g > (2n/\delta)\|g\|_p \} \le 2\varepsilon.$$

Hence, if we define

$$C(\alpha) = \sup \mathfrak{m}\{ T^* g > \alpha \|g\|_p \},$$

the function $C(\alpha)$ satisfies $\lim_{\alpha \to +\infty} C(\alpha) = 0$. □

This principle is completed with the fact that under the same hypothesis about (T_n), the set of $f \in \mathcal{L}^p[-\pi, \pi]$, where the limit $\lim_n T_n f(x)$ exists a. e., is closed in $\mathcal{L}^p[-\pi, \pi]$. To prove this, define the operator

$$\Omega(f)(x) = \limsup_{n,m} |T_n f(x) - T_m f(x)|.$$

It is clear that $\Omega f \leq 2T^* f$. Therefore

$$\mathfrak{m}\{\Omega f(x) > \alpha \|f\|_p\} \leq C(\alpha/2).$$

For every function φ such that the limit $\lim_n T_n \varphi(x)$ exists a. e., we have $\Omega \varphi = 0$, and $\Omega(f - \varphi) = \Omega f$. It follows that

$$\mathfrak{m}\{\Omega f(x) > \alpha \|f - \varphi\|_p\} \leq C(\alpha/2).$$

Now let f be in the closure of the sets of functions φ. Take $\alpha = 1/\varepsilon$ and $\|f - \varphi\|_p < \varepsilon^2$. We obtain

$$\mathfrak{m}\{\Omega f(x) > \varepsilon\} \leq C(1/2\varepsilon).$$

It follows easily that $\mathfrak{m}\{\Omega f(x) > 0\} = 0$.

2.5 Summability

As we have said, Du Bois Reymond constructed a continuous function whose Fourier series diverges at some point. Lipot Fejér proved, when he was 19 years old, that in spite of this we can recover a continuous function from its Fourier series.

Recall that if a sequence converges, there converges also, and to the same limit, the series formed by the arithmetic means of his terms. Fejér considered the mean values of the partial sums

$$\sigma_n(f, x) = \frac{1}{n+1} \sum_{j=0}^{n} S_n(f, x).$$

We have an integral expression for these mean values

$$\sigma_n(f, x) = \frac{1}{2\pi} \int_{-\pi}^{\pi} F_n(x - t) f(t) \, dt$$

where F_n is Fejér kernel:

$$F_n(t) = \frac{1}{n+1} \sum_{j=0}^{n} D_j(t) = \sum_{j=-n}^{j=n} \left(1 - \frac{|j|}{n+1}\right) e^{ijt}.$$

There is another expression for F_n. We substitute the value of the Dirichlet kernel, then we want to sum the sequence $\sin(n+1/2)t$. This is the imaginary part of

$$\sum_{j=0}^{n} e^{i(2j+1)t/2} = e^{-it/2} \sum_{j=0}^{n} e^{ijt} = \frac{1}{2i} \frac{e^{i(n+1)t} - 1}{\sin(t/2)}$$

thus

$$\sum_{j=0}^{n} \sin \frac{2j+1}{2} t = \frac{1 - \cos(n+1)t}{2\sin(t/2)} = \frac{\sin^2(n+1)\,{}^t\!/\!_2}{\sin(t/2)};$$

it follows that

$$F_n(t) = \frac{1}{n+1}\left(\frac{\sin(n+1)\,{}^t\!/\!_2}{\sin(t/2)}\right)^2. \tag{2.4}$$

We thus have that F_n is a positive function, $\|F_n\|_1 = 1$, and for every $\delta > 0$ we have, uniformly on $\delta < |t| \le \pi$, that $\lim_n F_n(t) = 0$.

With more generality we define a **summability kernel** to be a sequence (k_n) of periodic functions such that:

(i)
$$\frac{1}{2\pi}\int_{-\pi}^{\pi} k_n(t)\,dt = 1.$$

(ii)
$$\|k_n\|_1 \le C.$$

For every $\delta > 0$

(iii)
$$\lim_{n\to+\infty} \frac{1}{2\pi}\int_{\delta<|t|\le\pi} |k_n(t)|\,dt = 0.$$

In the following theorem we identify $[-\pi, \pi]$ with the torus $\mathbf{T}$ so that we can speak of $k_n * f$. Here is the theorem of Fejér, extended for every summability kernel.

Theorem 2.8 *Let (k_n) be a summability kernel. If $f : \mathbf{R} \to \mathbf{C}$ is continuous and 2π-periodic, then $k_n * f(x)$ converges uniformly to $f(x)$. Moreover, for every $1 \le p < +\infty$ and $f \in \mathcal{L}^p[-\pi, \pi]$ we have*

$$\lim_{n\to+\infty} \|k_n * f - f\|_p = 0.$$

Proof. First assume f to be continuous and 2π-periodic. By the property (i) of the summability kernel

$$k_n * f(x) - f(x) = \frac{1}{2\pi}\int_{-\pi}^{\pi} \big(f(x-t) - f(x)\big) k_n(t)\,dt.$$

Given $\varepsilon > 0$ we decompose the integral into two parts, one over $\{|t| < \delta\}$ and the other over $\{\delta < |t| < \pi\}$. The first is small by the continuity of f and property (ii) of the kernel; the second is small by (iii).

Observe that the same proof shows us the convergence at every point of continuity of f, for a measurable bounded f.

Since (F_n) is a summability kernel and $F_n * f$ is a trigonometrical polynomial for every f, it follows that these polynomials are dense on $\mathcal{C}(\mathbf{T})$.

Now for a $f \in \mathcal{L}^p[-\pi, \pi]$ we have that $t \mapsto G(t) = \|f(\cdot + t) - f(\cdot)\|_p$ is a continuous 2π-periodic function. We have

$$\|k_n * f - f\|_p \leq \frac{1}{2\pi} \int_{-\pi}^{\pi} \|f(\cdot - t) - f(\cdot)\|_p \, k_n(t) \, dt = k_n * G(0).$$

We can apply to this convolution the first part of the theorem to conclude that $\lim_n k_n * G(0) \to 0$.   □

Thus if f is continuous and 2π-periodic, $\sigma_n(f, x)$ converges uniformly to f, and converges to f in $\mathcal{L}^p[-\pi, \pi]$ if $f \in \mathcal{L}^p[-\pi, \pi]$. Another important example is that of the **Poisson kernel**. This kernel appear when we consider the Fourier series of f as the boundary values of a complex function defined on the open unit disc. If $f \in \mathcal{L}^1[-\pi, \pi]$ the series

$$\sum_{j=0}^{+\infty} \hat{f}(j) z^j + \sum_{j=1}^{+\infty} \hat{f}(-j) \bar{z}^j$$

converges on the open unit disc and defines a complex harmonic function $u(z)$. Then

$$u(re^{i\theta}) = \sum_{j=-\infty}^{+\infty} \hat{f}(j) r^{|j|} e^{ij\theta} = \frac{1}{2\pi} \int_{-\pi}^{\pi} P_r(\theta - t) f(t) \, dt,$$

where the Poisson kernel $P_r(\theta)$ is defined as

$$P_r(\theta) = \sum_{j=-\infty}^{+\infty} r^{|j|} e^{ij\theta} = \frac{1 - r^2}{1 - 2r \cos\theta + r^2}. \tag{2.5}$$

It is easy to check that $P_r(\theta)$ is a summability kernel. (Here the variable r takes the role of n, but this is a minor difference).

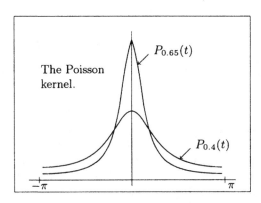

The Poisson kernel. $P_{0.65}(t)$ $P_{0.4}(t)$ $-\pi$ π

Then we have

Proposition 2.9 *If* $f \in \mathcal{L}^p[-\pi, \pi]$, $1 \le p < +\infty$, *or* $p = +\infty$ *and* f *is continuous with* $f(\pi) = f(-\pi)$, *we have* $\lim_{r \to 1^-} \|P_r * f - f\|_p = 0$.

Now we consider $f \in \mathcal{L}^1[-\pi, \pi]$ and ask about the a. e. convergence of $F_n * f(x)$ or $P_r * f(x)$ to $f(x)$. Obviously there exist some subsequences $F_{n_k} * f$ and $P_{r_k} * f$ that converge a. e. to f.

$u(re^{i\theta}) = P_r * f(\theta)$ is a harmonic function on the unit disc. Then what we want is a theorem of radial convergence of a harmonic function to $\lim_{r \to 1^-} u(re^{i\theta})$. The first theorem of this type is due to Fatou in 1905: A bounded and analytic function on the open unit disc has radial limits at almost every point of the boundary. We prefer to give a proof like that of the differentiation theorem that also can be extended to the case of $\sigma_n(f, x)$.

Theorem 2.10 (Fatou) *Let* $f \in \mathcal{L}^1[-\pi, \pi]$. *For almost every point* $x \in [-\pi, \pi]$ *we have*

$$\lim_{r \to 1^-} P_r * f(x) = f(x), \qquad \lim_n \sigma_n(f, x) = f(x).$$

Proof. The principal part is to prove that the maximal operators

$$P^* f(x) = \sup_{0 < r < 1} |P_r * f(x)|, \qquad F^* f(x) = \sup_n |F_n * f(x)|$$

are bounded. This follows from the general inequality about the maximal Hardy-Littlewood function. Define $f^\circ \colon \mathbf{R} \to \mathbf{C}$ as 0 for $|x| > 2\pi$ and equal to the periodic extension of f when $|x| < 2\pi$. Also put $P_r^\circ \colon \mathbf{R} \to \mathbf{C}$ as $P_r^\circ(\theta) = 0$ when $|\theta| > \pi$, and $P_r(\theta)$ for $|\theta| < \pi$. Then we can write

$$P_r * f(x) = P_r^\circ * f^\circ(x), \qquad |x| < \pi.$$

In the same way we can define the function F_n° so that

$$\sigma_n(f, x) = F_n * f(x) = F_n^\circ * f^\circ(x), \qquad |x| < \pi.$$

Since P_r° is a radial function that is decreasing for $x > 0$ and its integral on $\mathbf{R}$ is equal to 1, we have $|P_r^\circ * f^\circ(x)| \le \mathcal{M}f^\circ(x)$. Thus $P^* f(x) \le \mathcal{M}f^\circ(x)$ for every $|x| < \pi$.

The Fejér kernel is not decreasing, but $\sin(t/2) > t/\pi$ for $0 < t < \pi$, therefore

$$F_n^\circ(t) \le \begin{cases} \frac{1}{n+1}\left(\frac{(n+1)t/2}{t/4}\right)^2 \le 16(n+1), & \text{if } |t| \le \frac{1}{n+1}, \\ \frac{1}{n+1}\left(\frac{1}{t/4}\right)^2 = \frac{16}{n+1}\frac{1}{t^2}, & \text{if } \frac{1}{n+1} < |t| < \pi. \end{cases}$$

Thus F_n is bounded by a radial function that is decreasing for $t > 0$ and has integrals uniformly bounded. It follows that

$$F^* f(x) \leq C \mathcal{M} f(x).$$

Now the proof of the a. e. pointwise convergence follows as in the differentiability theorem.   □

2.6 The conjugate function

Since (e^{int}) is a complete orthonormal system on the space $\mathcal{L}^2[-\pi, \pi]$, we have Parseval equality

$$\sum_{j \in \mathbf{Z}} |\widehat{f}(j)|^2 = \frac{1}{2\pi} \int_{-\pi}^{\pi} |f(t)|^2 \, dt.$$

It follows that

$$\lim_n \|f - S_n(f)\|_2 = 0.$$

In fact for every $1 < p < +\infty$ we have

$$\lim_n \|f - S_n(f)\|_p = 0, \qquad (2.6)$$

for every $f \in \mathcal{L}^p[-\pi, \pi]$. In the case $p = 1$ this is no longer true. In fact if F_n and D_n are Fejér's and Dirichlet's kernels, then $S_n(F_N) = D_n * F_N = \sigma_N(D_n)$. Therefore by Fejér's Theorem $\lim_N S_n(F_N) = D_n$ on $\mathcal{L}^1[-\pi, \pi]$. Since $\|F_N\|_1 = 1$, it follows that $\|S_n\|_1 \geq L_n$. And we know that $L_n \sim \log n$.

On the other hand to prove (2.6), it suffices to prove that the norm of the operators of partial sums $S_n : \mathcal{L}^p[-\pi, \pi] \to \mathcal{L}^p[-\pi, \pi]$ is uniformly bounded. In fact, since the polynomials are dense on $\mathcal{L}^p[-\pi, \pi]$, given $\varepsilon > 0$ we find a polynomial P_ε such that $\|f - P_\varepsilon\|_p < \varepsilon$. Then if n is greater than the degree of P_ε

$$\|f - S_n(f)\|_p \leq \|f - P_\varepsilon\|_p + \|S_n(P_\varepsilon) - S_n(f)\|_p \leq \varepsilon + C\varepsilon.$$

The uniform boundedness of these norms was proved by M. Riesz in 1928. He considered the operator defined on the space of trigonometrical polynomials as

$$\mathcal{R}\left(\sum_j a_j e^{ijt}\right) = \sum_{j \geq 0} a_j e^{ijt}.$$

It is clear that $\mathcal{R}$ is a continuous projection on $\mathcal{L}^2[-\pi, \pi]$. What is remarkable about $\mathcal{R}$ is that we have the relationship

$$S_n(f) = e^{-int} \mathcal{R}(e^{int} f) - e^{i(n+1)t} \mathcal{R}(e^{-i(n+1)t} f).$$

Then the uniform boundedness of the norm of S_n follows if $\mathcal{R}$ can be extended to a continuous operator on $\mathcal{L}^p[-\pi, \pi]$.

The operator $\mathcal{R}$ is related to the conjugate harmonic function. Consider a power series
$$\sum_{n>0}(a_n + ib_n)z^n.$$

Its real and imaginary part for $z = e^{it}$ are
$$u = \sum_{n>0}(a_n \cos nt - b_n \sin nt); \qquad v = \sum_{n>0}(a_n \sin nt + b_n \cos nt).$$

We say that $v = \tilde{u}$ is the conjugate series to u. The operator $\mathcal{H}$ that sends u to v must satisfy
$$\mathcal{H}(\cos nt) = \sin nt; \qquad \mathcal{H}(\sin nt) = -\cos nt.$$

It is the same to say
$$\mathcal{H}(e^{int}) = -i\,\mathrm{sgn}(n)e^{int}.$$

The $\mathcal{R}$ and $\mathcal{H}$ operators are related by
$$\mathcal{R}\big(f(\theta)\big) + e^{i\theta}\mathcal{R}\big(e^{-i\theta}f(\theta)\big) = f(\theta) + i\mathcal{H}\big(f(\theta)\big).$$

It follows that the operator $\mathcal{H}$ extends to a continuous operator from $\mathcal{L}^2[-\pi, \pi]$ to $\mathcal{L}^2[-\pi, \pi]$.

In the next chapter we shall study the operator $\mathcal{H}$. For the time being we obtain some expression for this operator.

Let $f \in \mathcal{L}^1[-\pi, \pi]$. Its Fourier series is
$$\sum_j \widehat{f}(j)e^{ijx}.$$

We shall call
$$\sum_j (-i)\,\mathrm{sgn}(j)\widehat{f}(j)e^{ijx}$$

the conjugate series.

It is clear that when $f \in \mathcal{L}^2[-\pi, \pi]$, this conjugate series is the Fourier series of $\mathcal{H}f$.

We can express the partial sums of the conjugate series as a convolution
$$\tilde{S}_n(f, x) = \sum_{j=-n}^{n}(-i)\,\mathrm{sgn}(j)\widehat{f}(j)e^{ijx} = \frac{1}{2\pi}\int_{-\pi}^{\pi} f(t)\tilde{D}_n(x - t)\,dt.$$

Here $\tilde{D}_n$ is the conjugate of the Dirichlet kernel
$$\tilde{D}_n(t) = 2\sum_{j=1}^{n}\sin jt = \frac{\cos t/2 - \cos(n + 1/2)t}{\sin t/2}.$$

And we have a condition of convergence similar to that of Dini.

Theorem 2.11 (Pringsheim convergence test) *Let $f \in \mathcal{L}^1[-\pi, \pi]$ be a 2π-periodic function and $x \in [-\pi, \pi]$ such that*

$$\int_0^\pi |f(x+t) - f(x-t)| \frac{dt}{t} < +\infty,$$

then the conjugate series converges at the point x.

Proof. Since $\tilde{D}_n(t)$ is an odd function,

$$\tilde{S}_n(f, x) = \frac{1}{2\pi} \int_0^\pi \big(f(x-t) - f(x+t)\big) \tilde{D}_n(t)\, dt.$$

Then the Riemann-Lebesgue lemma and the expression for $\tilde{D}_n(t)$ end the proof.   □

We also get that

$$\lim \tilde{S}_n(f, x) = \frac{1}{2\pi} \int_0^\pi \frac{f(x-t) - f(x+t)}{\tan t/2}\, dt.$$

It follows easily that under the hypothesis of the theorem, there exists the principal value and

$$\lim \tilde{S}_n(f, x) = \frac{1}{2\pi}\, \mathrm{p.v.} \int_{-\pi}^\pi \frac{f(x-t)}{\tan t/2}\, dt.$$

We see that the Hilbert transform of a differentiable function is given by

$$\mathcal{H}f(x) = \frac{1}{2\pi}\, \mathrm{p.v.} \int_{-\pi}^\pi \frac{f(x-t)}{\tan t/2}\, dt.$$

In 1913 Luzin proved that the principal value exists and equals $\mathcal{H}f(x)$ a. e. for every $f \in \mathcal{L}^2[-\pi, \pi]$. Later, in 1919, Privalov proved that the principal value exists a. e. for every $f \in \mathcal{L}^1[-\pi, \pi]$.

2.7 The Hilbert transform on R

In the following chapter we will study the Hilbert transform. It is convenient perform this study on **R** instead of on the torus. Almost all of what we have said for Fourier series has an analogue for **R**.

For $f \in \mathcal{L}^1(\mathbf{R})$ the Fourier transform is defined as

$$\widehat{f}(x) = \int_{-\infty}^{+\infty} f(t) e^{-2\pi itx}\, dt.$$

This is analogous to the Fourier coefficients $\widehat{f}(j)$. The partial sums of the Fourier series are similar to

$$S_a(f,x) = \int_{-a}^{a} \widehat{f}(\xi)e^{2\pi i\xi x}\,d\xi = \int f(t)D_a(x-t)\,dt,$$

where the Dirichlet kernel is replaced by

$$D_a(t) = \frac{\sin 2\pi at}{\pi t}.$$

And Fejér sums are replaced by

$$\sigma_a(f,x) = \frac{1}{a}\int_0^a S_t(f,x)\,dt = \int f(x-\xi)F_a(\xi)\,d\xi,$$

where the analogue of the Fejér kernel is

$$\frac{1}{a}\left(\frac{\sin \pi\xi a}{\pi\xi}\right)^2.$$

The role of the unit disc is taken by the semiplane $y > 0$ on $\mathbf{R}^2$. If $f \in \mathcal{L}^1(\mathbf{R})$, we define on this semiplane the analytic function

$$F(z) = \frac{1}{\pi i}\int_{-\infty}^{+\infty}\frac{f(t)}{t-z}\,dt.$$

When f is a real function the real and imaginary parts of F are given by

$$u(x,y) = \frac{1}{\pi}\int_{-\infty}^{+\infty}\frac{y}{(x-t)^2+y^2}f(t)\,dt;$$

$$v(x,y) = \frac{1}{\pi}\int_{-\infty}^{+\infty}\frac{x-t}{(x-t)^2+y^2}f(t)\,dt.$$

For a general f the functions u and v defined by these integrals are harmonic conjugate functions.

The Hilbert transform is defined as

$$\mathcal{H}f(x) = \frac{1}{\pi}\,\text{p.v.}\int_{-\infty}^{+\infty}\frac{f(t)}{x-t}\,dt.$$

The study of this transform is equivalent to that of the transform on the torus. In fact, given $f \in \mathcal{L}^1[-\pi,\pi]$, if we define $f^\circ\colon \mathbf{R} \to \mathbf{C}$ as 0 for $|x| > 2\pi$ and equal to the periodic extension of f when $|x| < 2\pi$; then for every $|x| < \pi$

$$\mathcal{H}f(x) = \frac{1}{2\pi}\,\text{p.v.}\int_{-\pi}^{\pi}\frac{f^\circ(x-t)}{\tan t/2}\,dt.$$

Therefore

$$\mathcal{H}f(x) = \frac{1}{2\pi} \int_{-\pi}^{\pi} f^{\circ}(x-t)\left(\frac{1}{\tan t/2} - \frac{2}{t}\right) dt + \frac{1}{\pi} \text{p.v.} \int_{-\infty}^{+\infty} \frac{f^{\circ}(x-t)}{t} dt$$
$$- \frac{1}{\pi} \int_{|t|>\pi} \frac{f^{\circ}(x-t)}{t}.$$

If we designate by $\mathcal{H}_R$ and $\mathcal{H}_T$ the two transforms we have

$$|\mathcal{H}_T f(x) - \mathcal{H}_R f^{\circ}(x)| \leq C\|f\|_1.$$

And results for either transform can be transferred to the other.

2.8 The conjecture of Luzin

When Luzin published his paper in 1913, he knew the result of Fatou: the Poisson integral

$$\frac{1}{2\pi} \int_{-\pi}^{\pi} \frac{1-r^2}{1-2r\cos(t-x)+r^2} f(t)\, dt$$

converges a. e. to $f(x)$ for every $f \in \mathcal{L}^1[-\pi,\pi]$. He also knew the F. Riesz and E. Fischer result: given $\sum_{n=1}^{\infty}(a_n^2 + b_n^2) < +\infty$ there exists a function $f \in \mathcal{L}^2[-\pi,\pi]$ with Fourier series $\sum_{n=1}^{\infty} a_n \cos nx + b_n \sin nx$. He deduced that with the same hypotheses, there exists also the conjugate function $g \in \mathcal{L}^2[-\pi,\pi]$ with Fourier series $\sum_{n=1}^{\infty} -b_n \cos nx + a_n \sin nx$.

Then the coefficients a_n and b_n have integral representations in terms of f and also in terms of g. The analytic function $\sum_{n=1}^{\infty}(a_n - ib_n)(re^{ix})^n$ has two representations;

$$\frac{1}{2\pi} \int_{-\pi}^{\pi} \frac{(1-r^2)f(t)}{1-2r\cos(t-x)+r^2}\, dt = \frac{1}{2\pi} \int_{-\pi}^{\pi} \frac{2rg(t)\sin(t-x)}{1-2r\cos(t-x)+r^2}\, dt.$$

Fatou's result gives then that the second integral also converges a. e. to $f(x)$ when $r \to 1^-$.

He proved then the following result:

Theorem 2.12 *Let $f \in \mathcal{L}^2[-\pi,\pi]$, then*

$$\lim_{r\to 1^-} \left(\frac{1}{2\pi} \int_{-\pi}^{\pi} \frac{2rg(t)\sin(t-x)}{1-2r\cos(t-x)+r^2}\, dt - \frac{1}{2\pi} \int_{\eta<|t|<\pi} \frac{g(x-t)}{\tan t/2}\, dt \right) = 0$$

a. e., where $\eta = \eta(r)$ is the unique solution of $\cos x = 2r/(1+r^2)$ on the interval $(0,\pi/2)$.

Therefore he had proved that for $f \in \mathcal{L}^2(\mathbf{R})$

$$\frac{1}{2\pi} \, \text{p.v.} \int_{-\pi}^{\pi} \frac{g(x-t)}{\tan t/2} \, dt = f(x), \qquad a. \ e.$$

Then he remarked that

$$S_n(f,x) = \sum_{j=1}^{n} a_j \cos jx + b_j \sin jx$$

$$= \frac{1}{2\pi} \int_{-\pi}^{\pi} g(x+t) \Big(\frac{1}{\tan t/2} - \frac{\cos(n+1/2)t}{\sin t/2} \Big) \, dt.$$

Therefore $\lim S_n(f,x) = f(x)$ a. e., if and only if

$$\lim_{n\to\infty} \text{p.v.} \int_{-\pi}^{\pi} g(x+t) \frac{\cos nt}{t} \, dt = 0, \qquad a. \ e. \tag{2.7}$$

Then he knew that for every $g \in \mathcal{L}^2(\mathbf{R})$ the principal value

$$\text{p.v.} \int \frac{g(x+t)}{t} \, dt \quad \text{exists a. e.} \tag{2.8}$$

He noticed that this is not due to the smallness of the integrand. In fact, he knew that there exists a continuous function f and a set of positive measure A so that

$$\int \Big| \frac{f(x+t) - f(x-t)}{t} \Big| \, dt = +\infty$$

for $x \in A$.

But the proof he knew of (2.8) used as we have seen the theory of functions of a complex variable. In this proof it is not clear how it is that the cancellation between the positive and negative values produces the existence of the principal value. He conjectured that in a more constructive proof this cancellation would be clear and would not be disturbed by the presence of the factor $\cos nt$ in (2.7). From these considerations he conjectured that every $f \in \mathcal{L}^2[-\pi,\pi]$ would have a Fourier series a. e. convergent, that is $S_n(f,x) \to f(x)$ a. e. on $[-\pi,\pi]$. Carleson proved this in 1966. Later Hunt proved that this is true for every $f \in \mathcal{L}^p[-\pi,\pi]$ with $1 < p \le +\infty$. We shall prove these theorems in the second part.

Real proofs that $\mathcal{H}f$ exists for $f \in \mathcal{L}^p(\mathbf{R})$ were given, first by Besikovitch in 1926 for $p = 2$, then by Titchmarsh in 1926 for $p > 1$ and finally by Besikovitch for $p = 1$. But these proofs did not satisfy Luzin, because they were very complicated. Nevertheless, it was the right procedure. The results and techniques developed by these authors are needed in the final proof of Carleson's result.

3. Hilbert Transform

3.1 Introduction

The properties of the Hilbert transform not only inspired Luzin's conjecture about Fourier series of functions on $\mathcal{L}^2$; they are also needed in the proof of the Carleson-Hunt Theorem.

We need the existence almost everywhere of the principal value integral

$$\mathcal{H}f(x) = \text{p.v.} \int \frac{f(t)}{x-t}\, dt = \lim_{\varepsilon \to 0+} \int_{|x-t|>\varepsilon} \frac{f(t)}{x-t}\, dt,$$

for every $f \in \mathcal{L}^1(\mathbf{R})$ and also the bound

$$\|\mathcal{H}^* f\|_p \leq C_p \|f\|_p,$$

where $\mathcal{H}^* f$ denotes the maximal operator

$$\mathcal{H}^* f(x) = \sup_{\varepsilon > 0} \left| \int_{|x-t|>\varepsilon} \frac{f(t)}{x-t}\, dt \right|.$$

To obtain the fine result of Sjölin: $S_n(f, x)$ converges a. e. when

$$\int_{-\pi}^{\pi} |f(t)| \log^+ |f(t)| \log^+ \log^+ |f(t)|\, dt < +\infty,$$

it is necessary to estimate the constant C_p in the previous inequality.

3.2 Trunctated operators on $\mathcal{L}^2(\mathbf{R})$

We begin studying the truncated operators

$$\mathcal{H}_\varepsilon f(x) = K_\varepsilon * f(x) = \int_{|x-t|>\varepsilon} \frac{f(t)}{x-t}\, dt.$$

Here K_ε denotes the function equal to $1/t$ for $|t| > \varepsilon$ and 0 otherwise. As $K_\varepsilon \in \mathcal{L}^p(\mathbf{R})$ for every $1 < p \leq +\infty$, the convolution is defined for every $f \in \mathcal{L}^p(\mathbf{R})$, $1 \leq p < +\infty$ and $K_\varepsilon * f$ is a continuous and bounded function.

We want to prove that the operator $\mathcal{H}_\varepsilon \colon \mathcal{L}^p(\mathbf{R}) \to \mathcal{L}^p(\mathbf{R})$ is bounded by a constant that does not depend on ε. To achieve this we apply the interpolation of operators.

First consider the case $p = 2$. The Fourier transform is an isometry of $\mathcal{L}^2(\mathbf{R})$. Also since $\mathcal{H}_\varepsilon f$ is a convolution we have $\widehat{\mathcal{H}_\varepsilon f} = \widehat{K_\varepsilon} \cdot \widehat{f}$. Hence the norm of $\mathcal{H}_\varepsilon$ is equal to the norm of the operator that sends $g \in \mathcal{L}^2$ to $\widehat{K_\varepsilon} g$. In particular K_ε is bounded if $\widehat{K_\varepsilon}$ is in $\mathcal{L}^\infty$, and $\|\mathcal{H}_\varepsilon\| = \|\widehat{K_\varepsilon}\|_\infty$.

Hence the result for $p = 2$ is reduced to the calculation of $\widehat{K_\varepsilon}$. Since K_ε is not integrable, we must calculate its transform as a limit of the functions

$$\widehat{K_{\varepsilon,R}}(x) = \int_{R > |t| > \varepsilon} e^{-2\pi i x t} \frac{dt}{t} = -2i \int_\varepsilon^R \frac{\sin 2\pi x t}{t} \, dt,$$

where the limit is taken in $\mathcal{L}^2(\mathbf{R})$. As the pointwise limit of these functions when $R \to +\infty$ exists, it is equal to the limit in $\mathcal{L}^2(\mathbf{R})$.

Hence

$$\widehat{K_\varepsilon}(x) = -2i \int_\varepsilon^{+\infty} \frac{\sin 2\pi x t}{t} \, dt = -2i \int_y^{+\infty} \frac{\sin t}{t} \, dt,$$

for $y = 2\pi x \varepsilon$, and it is easy to see that these integrals are uniformly bounded by an absolute constant.

3.3 Truncated operators on $\mathcal{L}^1(\mathbf{R})$

It is not true that $\mathcal{H}_\varepsilon$ maps $\mathcal{L}^1(\mathbf{R})$ on $\mathcal{L}^1(\mathbf{R})$. What can be proved is only a weak type inequality.

If $f \in \mathcal{L}^p(\mu)$, we have the relation $\mu\{x \in X : |f(x)| > t\} \leq \|f\|_p^p t^{-p}$. A function that satisfies an inequality of the type

$$\mu\{x \in X : |f(x)| > t\} \leq \frac{C^p}{t^p}$$

is not necessarily contained in $\mathcal{L}^p(\mu)$. We say it is in weak $\mathcal{L}^p$. The best constant $C = \|f\|_p^*$ that satisfies the above inequality is called the weak $\mathcal{L}^p$ norm of f. Also an operator T, defined on every $f \in \mathcal{L}^p(\mu)$, is said to be of type (p, q) if $\|Tf\|_q \leq C\|f\|_p$ and of weak type (p, q) if $\nu\{y \in Y : |Tf(y)| > t\} \leq C^q\|f\|_p^q/t^q$. It is the same to say that $\|Tf\|_q^* \leq C\|f\|_p$.

Then what we shall prove is that the operator $\mathcal{H}_\varepsilon$ is of weak type $(1,1)$. We give a proof that can be extended to more general operators. It is based on the so-called decomposition of Calderón-Zygmund.

Theorem 3.1 (Decomposition of Calderón-Zygmund) *Let $f \in \mathcal{L}^1(\mathbf{R})$ and a positive real number α be given. There exists a decomposition $f = g + b$ (a good and a bad function) with the following properties. There exists an open set $\Omega = \bigcup_j Q_j$ where Q_j are nonoverlapping open intervals such that $\int_{Q_j} b(t)\,dt = 0$ for every j. On $F = \mathbf{R} \smallsetminus \Omega$ the function g is bounded by α. On every Q_j the function g is constant and equal to the mean value of f on Q_j and*

$$\alpha \le \frac{1}{|Q_j|} \int_{Q_j} |f(t)|\,dt, \quad \left| \frac{1}{|Q_j|} \int_{Q_j} f(t)\,dt \right| \le 2\alpha.$$

Proof. We consider the line to be decomposed into disjoint intervals on which the mean value of $|f|$ is less than α. This can be achieved taking large intervals. Now we subdivide each of these intervals into two equal intervals. For each one we calculate the mean value of $|f|$. If one of these mean values is greater than α we take the corresponding interval as one of the Q_j. We continue the process dividing those intervals on which the mean value is less than α. Finally we set Ω equal to the union of those intervals on which the mean value is greater than α. Then we set

$$b(x) = f(x)\,\chi_\Omega(x) - \sum_j \left(\frac{1}{|Q_j|} \int_{Q_j} f(t)\,dt \right) \chi_{Q_j}(x)$$

$$g(x) = f(x)\,\chi_F(x) + \sum_j \left(\frac{1}{|Q_j|} \int_{Q_j} f(t)\,dt \right) \chi_{Q_j}(x).$$

Now it is easy to prove that these functions satisfy all our conditions. The Q_j are disjoint by construction. For almost every point $x \in F$, the complement of Ω, there is a sequence of intervals J_n, with $x \in J_n$ and such that the mean value of $|f|$ on every one is less than α. By the differentiation theorem the value of $f(x)$ is also less than α if x is a point of Lebesgue of f.

Every Q_j is the half of an interval J where the mean value of $|f|$ is less than α. Therefore

$$\left| \frac{1}{|Q_j|} \int_{Q_j} f(t)\,dt \right| \le \frac{2}{|J|} \int_J |f(t)|\,dt \le 2\alpha.$$

$\square$

With this decomposition we can prove:

Theorem 3.2 (Kolmogorov) *For every $f \in \mathcal{L}^1(\mathbf{R})$ and $\alpha > 0$*

$$\mathfrak{m}\{x \in \mathbf{R} : |\mathcal{H}_\varepsilon f(x)| > \alpha\} \le C \frac{\|f\|_1}{\alpha}. \tag{3.1}$$

Proof. Let $f = g + b$ be the Calderón-Zygmund decomposition at the level α, and also let $\Omega = \bigcup_j Q_j$ the corresponding open set. We know that g and b are in $\mathcal{L}^1$, since

$$\|g\|_1 = \int_F |f| \, dm + \sum_j \frac{1}{|Q_j|} \left| \int_{Q_j} f \, dm \right| m(Q_j) \le \|f\|_1.$$

Therefore

$$\{x \in \mathbf{R} : \mathcal{H}_\varepsilon f > 2\alpha\} \subset \{x \in \mathbf{R} : \mathcal{H}_\varepsilon g > \alpha\} \cup \{x \in \mathbf{R} : \mathcal{H}_\varepsilon b > \alpha\}.$$

But g is also in $\mathcal{L}^2$, and since $|g| \le \alpha$ on F and $\le 2\alpha$ on Ω

$$\|g\|_2^2 \le \alpha \int_F |f| \, dm + 2\alpha \int_\Omega |f| \, dm \le 2\alpha \|f\|_1.$$

It follows that

$$m\{x \in \mathbf{R} : \mathcal{H}_\varepsilon g > \alpha\} \le \frac{\|\mathcal{H}_\varepsilon g\|_2^2}{\alpha^2} \le C \frac{2\alpha \|f\|_1}{\alpha^2} \le A \frac{\|f\|_1}{\alpha}.$$

Now we start with the bad function. First observe that if $G = \bigcup_j 2Q_j$, we have

$$m(G) \le 2 \sum_j m(Q_j) \le 2 \sum_j \frac{1}{\alpha} \int_{Q_j} |f| \, dm \le \frac{2}{\alpha} \|f\|_1.$$

Therefore we have

$$m\{x \in \mathbf{R} : \mathcal{H}_\varepsilon b(x) > \alpha\} \le m(G) + m\{x \in \mathbf{R} \smallsetminus G : \mathcal{H}_\varepsilon b(x) > \alpha\}.$$

To obtain the corresponding inequality we calculate

$$\int_{\mathbf{R} \smallsetminus G} |\mathcal{H}_\varepsilon b(x)| \, dx \le \sum_j \int_{\mathbf{R} \smallsetminus G} \left| \int_{Q_j} K_\varepsilon(x - t) b(t) \, dt \right| dx$$

where we have used that b is zero on F. Now applying that the integral of b is zero on every Q_j, we get

$$\int_{\mathbf{R} \smallsetminus G} |\mathcal{H}_\varepsilon b(x)| \, dx \le \sum_j \int_{\mathbf{R} \smallsetminus G} \left| \int_{Q_j} (K_\varepsilon(x - t) - K_\varepsilon(x - t_j)) b(t) \, dt \right| dx$$

where t_j denotes the center of Q_j. Thus, by Fubini's theorem, the last integral is less than

$$\sum_j \int_{Q_j} \left(\int_{|x - t_j| > 2|t - t_j|} |K_\varepsilon(x - t) - K_\varepsilon(x - t_j)| \, dx \right) |b(t)| \, dt$$

Now we claim: the integral on x is bounded by an absolute constant B. Therefore

$$\int_{\mathbf{R}\smallsetminus G} |\mathcal{H}_\varepsilon b(x)|\, dx \le B \sum_j \int_{Q_j} |b(t)|\, dt \le 2B\|f\|_1.$$

Consequently

$$\mathrm{m}\{x \in \mathbf{R} \smallsetminus G : \mathcal{H}_\varepsilon b(x) > \alpha\} \le \frac{2B\|f\|_1}{\alpha}.$$

We only have to collect the results obtained. We have to prove the claim. This will be done in the following proposition. □

In the following it will be convenient to use Iverson-Knuth's notation: If $P(x)$ is a condition that can be true or false for every x, then $[P(x)]$ by definition, is equal to 1 if $P(x)$ is true and 0 if $P(x)$ is false. In other words, $[P(x)]$ is the characteristic function of the set $\{x \in \mathbf{R} : P(x)\}$.

The claim will be a consequence of the following fact:

Proposition 3.3 *For every $a \in \mathbf{R}$ and $\varepsilon > 0$ there exists an even function $\psi : \mathbf{R} \to [0, +\infty)$ such that it is decreasing on $[0, +\infty)$,*

$$\big|K_\varepsilon(t+a) - K_\varepsilon(t)\big|[|t| > 2|a|] \le \psi(x) \quad \text{for every } |t| \ge |x|,$$

and the integral is bounded by an absolute constant $\int \psi(t)\, dt < C$.

Proof. We have

$$\big|K_\varepsilon(t+a) - K_\varepsilon(t)\big| \cdot [|t| > 2|a|] \le$$
$$\big|K(t+a) - K(t)\big| \cdot [|t+a| > \varepsilon] \cdot [|t| > 2|a|]$$
$$+ \big|K(t)\big| \cdot \big|[|t+a| > \varepsilon] - [|t| > \varepsilon]\big| \cdot [|t| > 2|a|].$$

That is, for $\varepsilon < 3|a|$

$$\le \frac{|a|}{|t|(|t| - |a|)}[|t| > 2|a|] + \frac{1}{|t|}[2|a| \le |t| \le 4|a|].$$

And for $\varepsilon \ge 3|a|$

$$\le \frac{|a|}{|t|(|t| - |a|)}[|t| > 2|a|] + \frac{1}{|t|}\Big[\frac{2}{3}\varepsilon \le |t| \le \frac{4}{3}\varepsilon\Big].$$

Now let

$$\psi_1(a, t) = \begin{cases} |a|/(|t|(|t| - |a|)) & \text{if } |t| > 2|a|, \\ 1/2|a| & \text{if } |t| \le 2|a|. \end{cases}$$

$$\psi_2(\varepsilon, t) = \begin{cases} |t|^{-1} & \text{if } 2\varepsilon/3 \le |t| \le 4\varepsilon/3, \\ \dfrac{3}{2\varepsilon} & \text{if } |t| \le 2\varepsilon/3, \\ 0 & \text{if } |t| > 4\varepsilon/3. \end{cases}$$

Then we can take $\psi(t) = \psi_1(a,t) + \psi_2(3|a|,t)$ in case $\varepsilon < 3|a|$ and $\psi(t) = \psi_1(a,t) + \psi_2(\varepsilon,t)$ when $\varepsilon \geq 3|a|$.

It is clear that this function satisfies all our conditions.   □

This proposition will be needed in the proof of Cotlar's Inequality (Theorem 3.7)

3.4 Interpolation

Assume that T is a linear operator defined on some space of measurable functions that contains $\mathcal{L}^{p_0}(\mu)$ and $\mathcal{L}^{p_1}(\mu)$ for some $1 \leq p_1 < p_0 \leq +\infty$. Then T is defined also on every $p \in [p_1, p_0]$. This is true because we can decompose every function $f \in \mathcal{L}^p(\mu)$

$$f = f_0 + f_1,$$

where, being $A = \{t \in X : |f(t)| < 1\}$, we define

$$f_0 = f\chi_A \quad \text{and} \quad f_1 = f - f_0.$$

Then $|f_0| \leq 1$ and $|f_0| \leq |f|$, and so $f_0 \in \mathcal{L}^\infty(\mu)$ and $f_0 \in \mathcal{L}^p(\mu)$. This implies that $f_0 \in \mathcal{L}^{p_0}(\mu)$. In an analogous way $|f_1| \leq |f|$ and $|f_1| \leq (|f|)^p$, therefore $f_1 \in \mathcal{L}^p(\mu)$ and $f_1 \in \mathcal{L}^1(\mu)$, and so it is in the intermediate $\mathcal{L}^{p_1}(\mu)$. Then it is clear that $T(f) = T(f_0) + T(f_1)$ is well defined.

The interpolation theorem gives a quantitative version of this observation. We will prove the Marcinkiewicz interpolation theorem in chapter 11. We apply now this theorem. The reader can read this chapter now or even only the proof of theorem 11.10 and some previous definitions needed in it.

Proposition 3.4 *For every $1 < p < +\infty$ the operator $\mathcal{H}_\varepsilon : \mathcal{L}^p(R) \to \mathcal{L}^p(\mathbf{R})$ is continuous and $\|\mathcal{H}_\varepsilon\|_p \leq Cp^2/(p-1)$.*

Proof. We have proved that $\mathcal{H}_\varepsilon$ is of weak type $(1,1)$ and strong type $(2,2)$. Therefore applying Marcinkiewicz's Theorem 11.10 for $1 < p \leq 2$ we obtain

$$\|\mathcal{H}_\varepsilon f\|_p \leq \frac{Cp}{(p-1)(2-p)} \|f\|_p.$$

Values of $p > 2$ are conjugates to $p' < 2$. And it is easy to see that $\|\mathcal{H}_\varepsilon\|_p = \|\mathcal{H}_\varepsilon\|_{p'}$. This follows from Fubini's theorem: If $f \in \mathcal{L}^{p'}$ and $g \in \mathcal{L}^p$ we have

$$\int_{\mathbf{R}} \left(\int_{|x-t|>\varepsilon} \frac{f(t)}{x-t}\, dt \right) g(x)\, dx = \int_{\mathbf{R}} \left(\int_{|x-t|>\varepsilon} \frac{g(x)}{x-t}\, dx \right) f(t)\, dt$$

This implies that if $|p - 2| > 1/2$, then $\|\mathcal{H}_\varepsilon\|_p \leq Cp^2/(p - 1)$. (This is true for $1 < p < 3/2$ and the bound can be written in the symmetric form Cpp')

We only have to prove that the norm $\|\mathcal{H}_\varepsilon\|_p$ is uniformly bounded for $|p - 2| \leq 1/2$, and this can be done by applying again the interpolation theorem, this time between $p_1 = 4/3$ and $p_0 = 4$. $\qquad\square$

3.5 The Hilbert transform

Now we are in position to prove that, for every $1 < p < +\infty$ there is a bounded operator $\mathcal{H}: \mathcal{L}^p(\mathbf{R}) \to \mathcal{L}^p(\mathbf{R})$. We shall prove that if $1 < p < +\infty$ and $f \in \mathcal{L}^p(\mathbf{R})$, then there exists the limit $\lim_{\varepsilon \to 0+} \mathcal{H}_\varepsilon f$ (being taken in the space $\mathcal{L}^p(\mathbf{R})$).

In the case $p = 1$, $\mathcal{H}_\varepsilon f$ is only in weak $\mathcal{L}^1$ so that we have to modify slightly the reasoning.

We shall need a dense subset where the limits exist.

Proposition 3.5 *Let φ be an infinitely differentiable function of compact support. For every $1 < p < +\infty$ the limit*

$$\mathcal{H}\varphi = \lim_{\varepsilon \to 0+} \mathcal{H}_\varepsilon\varphi$$

exists on $\mathcal{L}^p(\mathbf{R})$. Moreover for every x there exists the limit

$$\mathcal{H}\varphi(x) = \lim_{\varepsilon \to 0+} \mathcal{H}_\varepsilon\varphi(x).$$

Proof. Observe that for $0 < \delta < \varepsilon$

$$\mathcal{H}_\varepsilon\varphi(x) - \mathcal{H}_\delta\varphi(x) = \int_{\delta < |t| < \varepsilon} \frac{1}{t}\varphi(x - t)\,dt = \int_{\delta < |t| < \varepsilon} \frac{1}{t}\left(\varphi(x - t) - \varphi(x)\right)dt.$$

Therefore

$$\|\mathcal{H}_\varepsilon\varphi - \mathcal{H}_\delta\varphi\|_p \leq \int_{\delta < |t| < \varepsilon} \frac{1}{t}\|\varphi(\cdot - t) - \varphi(\cdot)\|_p\,dt.$$

The hypothesis about φ implies that $\|\varphi(\cdot - t) - \varphi(\cdot)\|_p \leq C|t|$. It follows that

$$\|\mathcal{H}_\varepsilon\varphi - \mathcal{H}_\delta\varphi\|_p \leq C(\varepsilon - \delta).$$

We can think of $\mathcal{H}_\varepsilon\varphi$ as a convolution of an $\mathcal{L}^1$ with an $\mathcal{L}^p$ function ($p > 1$); hence it is an $\mathcal{L}^p$ function.

We apply the completeness of $\mathcal{L}^p(\mathbf{R})$ to get our first assertion.

The conclusion about pointwise convergence can be derived in the same way, since $\|\varphi(\cdot - t) - \varphi(\cdot)\|_\infty \leq C|t|$. $\qquad\square$

Observe that under the same hypotheses about φ, we can prove in the case $p = 1$ the inequality $\|\mathcal{H}_\varepsilon\varphi - \mathcal{H}_\delta\varphi\|_1 \leq C(\varepsilon - \delta)$, but $\mathcal{H}_\varepsilon\varphi \notin \mathcal{L}^1(\mathbf{R})$.

Now we can prove that we can define for every $f \in \mathcal{L}^p(\mathbf{R})$

$$\mathcal{H}f = \lim_{\varepsilon \to 0^+} \mathcal{H}_\varepsilon f,$$

where the limits must be understood in the sense of the norm of $\mathcal{L}^p(\mathbf{R})$, $p > 1$. In fact,

$$\|\mathcal{H}_\varepsilon f - \mathcal{H}_\delta f\|_p \leq \|\mathcal{H}_\varepsilon\varphi - \mathcal{H}_\delta\varphi\|_p + \|\mathcal{H}_\varepsilon(f - \varphi)\|_p + \|\mathcal{H}_\delta(f - \varphi)\|_p.$$

Therefore we can prove for $p > 1$, given the density of the smooth functions, that $\mathcal{H}_\varepsilon f$ satisfies the Cauchy condition of convergence. The problem in the case $p = 1$ is that we only know the weak inequality given by Kolmogorov theorem above.

Almost all the reasoning for $p > 1$ can be implemented for $p = 1$. We shall consider the space $\mathcal{L}^{1,\infty}(\mathbf{R})$ of all the measurable functions $f : \mathbf{R} \to \mathbf{C}$ such that $\|f\|_{1,\infty} < +\infty$. This is defined as

$$\|f\|_{1,\infty} = \sup_{t>0} t \cdot \mathfrak{m}\{x \in \mathbf{R} : |f(x)| > t\}.$$

This is not a norm but satisfies

$$\|af\|_{1,\infty} = |a| \cdot \|f\|_{1,\infty}, \qquad \|f + g\|_{1,\infty} \leq 2\|f\|_{1,\infty} + 2\|g\|_{1,\infty}.$$

Therefore $\mathcal{L}^{1,\infty}(\mathbf{R})$ is a vector space. Also for $f \in \mathcal{L}^1(\mathbf{R})$ we have $\|f\|_{1,\infty} \leq \|f\|_1$. Sometimes we call $\mathcal{L}^{1,\infty}(\mathbf{R})$ the **weak $\mathcal{L}^1$ space**.

This quasi-norm allows us to define a topology on $\mathcal{L}^{1,\infty}(\mathbf{R})$ where a basis of neighborhoods of f are the sets $f + B(0, \varepsilon)$, where $B(0, \varepsilon) = \{g \in \mathcal{L}^{1,\infty}(\mathbf{R}) : \|g\|_{1,\infty} < \varepsilon\}$.

Now for $f \in \mathcal{L}_1(\mathbf{R})$ and a smooth φ

$$\|\mathcal{H}_\varepsilon f - \mathcal{H}_\delta f\|_{1,\infty} \leq 3\|\mathcal{H}_\varepsilon\varphi - \mathcal{H}_\delta\varphi\|_{1,\infty} + 3\|\mathcal{H}_\varepsilon(f - \varphi)\|_{1,\infty}$$
$$+ 3\|\mathcal{H}_\delta(f - \varphi)\|_{1,\infty} \leq C_\varphi|\varepsilon - \delta| + C\|f - \varphi\|_1.$$

We are able to prove the *condition of Cauchy* for $\mathcal{H}_\varepsilon f$ in the space $\mathcal{L}^{1,\infty}(\mathbf{R})$. All that is needed now is to prove that the space $\mathcal{L}^{1,\infty}(\mathbf{R})$ is complete.

Proposition 3.6 *Let (f_n) be a Cauchy sequence in $\mathcal{L}^{1,\infty}(\mathbf{R})$. Then there exists a subsequence (f_{n_k}) and a measurable function $f \in \mathcal{L}^{1,\infty}(\mathbf{R})$, such that f_{n_k} converges a. e. to f, and $\lim_n \|f_n - f\|_{1,\infty} = 0$.*

Proof. The proof is the same as that of the Riesz-Fischer theorem. We select the subsequence $g_k = f_{n_k}$ in such way that $\|g_{k+1} - g_k\|_{1,\infty} < 4^{-k}$. If $Z_k = \{|g_{k+1} - g_k| > 2^{-k}\}$, we have $\mathfrak{m}(Z_k) \leq 2^k \cdot 4^{-k} = 2^{-k}$. Therefore $Z = \bigcap_N \bigcup_{k>N} Z_k$ is of measure zero and for every $x \notin Z$ there exists N such that for every $n > N$, $|g_{n+1}(x) - g_n(x)| \leq 2^{-n}$. It follows that the sequence (g_n) converges a. e. to some measurable function f.

To prove that $\lim_n \|f_n - f\|_{1,\infty} = 0$ it is convenient to observe that for every finite sequence of functions (h_j) in $\mathcal{L}^{1,\infty}(\mathbf{R})$ we have

$$\|\sum_{j=1}^{N} h_j\|_{1,\infty} \leq \sum_{j=1}^{N} 2^j \|h_j\|_{1,\infty}.$$

Therefore we have $\|g_{n+k} - g_n\|_{1,\infty} \leq 4^{1-n}$. That is,

$$m(A_k(\alpha)) = m\{|g_{n+k} - g_n| > \alpha\} \leq \frac{1}{4^{n-1}\alpha}.$$

Since $\{|f - g_n| > \alpha\} \subset Z \cup \bigcup_N \bigcap_{k=N}^{\infty} A_k(\alpha)$, it follows easily that

$$m\{|f - g_n| > \alpha\} \leq 4^{1-n}\alpha^{-1}.$$

Therefore $\|f - g_n\|_{1,\infty} \to 0$.   □

Now we can define the Hilbert transform $\mathcal{H}f$ for every $f \in \mathcal{L}^p(\mathbf{R})$. For $1 < p < +\infty$, $\mathcal{H}f \in \mathcal{L}^p(\mathbf{R})$. Also $\mathcal{H}: \mathcal{L}^p(\mathbf{R}) \to \mathcal{L}^p(\mathbf{R})$ is a bounded linear operator and

$$\|\mathcal{H}f\|_p \leq C\frac{p^2}{p-1}\|f\|_p, \qquad 1 < p < +\infty.$$

In the case $p = 1$ the linear operator $\mathcal{H}$ is defined but it takes values in $\mathcal{L}^{1,\infty}$. In particular for every $f \in \mathcal{L}^1(\mathbf{R})$, $\|\mathcal{H}f\|_{1,\infty} \leq C\|f\|_1$. This follows from the corresponding Kolmogorov theorem about $\mathcal{H}_\varepsilon$ and from the fact that for some sequence (ε_n) we have that $\mathcal{H}_{\varepsilon_n} f$ converges a. e. to $\mathcal{H}f$.

3.6 Maximal Hilbert transform

In the proof of Carleson's Theorem we need the inequality $\|\mathcal{H}^* f\|_p \leq Bp\|f\|_p$, valid for every $p \geq 2$ and every $f \in \mathcal{L}^p(\mathbf{R})$, and where

$$\mathcal{H}^* f(x) = \sup_{\varepsilon > 0} |\mathcal{H}_\varepsilon f(x)|$$

is the **maximal Hilbert transform**. Historically this result is obtained to give a **real** proof of the pointwise convergence

$$\mathcal{H}f(x) = \lim_{\varepsilon \to 0^+} \mathcal{H}_\varepsilon f(x), \qquad a.\ e.$$

The proof which we shall give of this result is based on the following bound.

Theorem 3.7 (Cotlar's Inequality) *Let $f \in \mathcal{L}^p(\mathbf{R})$, $1 < p < +\infty$. Then*

$$\mathcal{H}^* f(x) \leq A\big(\mathcal{M}(\mathcal{H}f)(x) + \mathcal{M}f(x)\big).$$

Therefore

$$\|\mathcal{H}^* f\|_p \leq Bp\|f\|_p, \qquad p \geq 2.$$

Proof. Fix $a \in \mathbf{R}$ and $\varepsilon > 0$. We want to prove

$$|\mathcal{H}_\varepsilon f(a)| \leq A\big(\mathcal{M}(\mathcal{H}f)(a) + \mathcal{M}f(a)\big).$$

We notice that

$$\mathcal{H}_\varepsilon f(a) = \int_{|t-a|>\varepsilon} \frac{f(t)}{a-t}\, dt = \int_{\mathbf{R}} \frac{f_2(t)}{a-t}\, dt = \mathcal{H}f_2(a),$$

where $f = f_1 + f_2$ and f_2 is equal to f on $|t - a| > \varepsilon$ and is equal to 0 on the interval $2J = \{t : |a - t| \leq \varepsilon\}$. This equality has a sense in spite of the fact that we have defined $\mathcal{H}$ only as an operator from $\mathcal{L}^p(R)$ to $\mathcal{L}^p(R)$. In fact, since f_2 is null on the interval $2J$, at every point x of the interval $J = \{t : |a - t| \leq \varepsilon/2\}$, we have that $\mathcal{H}_\eta f_2(x)$ does not depend on η for $\eta < \varepsilon/2$. Therefore $\mathcal{H}f_2$ is equal to $\mathcal{H}_\eta f_2$ on J and is a continuous function on this interval.

We can bound the oscillation of $\mathcal{H}f_2$ on J.

$$|\mathcal{H}f_2(x) - \mathcal{H}f_2(a)| \leq \int \Big|K_\eta(x - t) - K_\eta(a - t)\Big| \cdot |f_2(t)|\, dt.$$

Now we recall that there exists an even function $\psi(x)$ that decreases with $|x|$, its integral is bounded by an absolute constant, and such that

$$\Big|K_\eta(x - t) - K_\eta(a - t)\Big|[|a - t| > 2|x - a|] \leq \psi(a - t).$$

(cf. proposition 3.3). Therefore, it follows that

$$|\mathcal{H}f_2(x) - \mathcal{H}f_2(a)| \leq \int_{\mathbf{R}} \psi(a - t) f_2(t)\, dt \leq C\mathcal{M}f_2(a) \leq C\mathcal{M}f(a),$$

by the general inequality satisfied by the Hardy-Littlewood maximal function.

Collecting our results we have $|\mathcal{H}_\varepsilon f(a)| \leq C\mathcal{M}f(a) + |\mathcal{H}f_2(x)|$, for every $x \in J$. Therefore

$$|\mathcal{H}_\varepsilon f(a)| \leq C\mathcal{M}f(a) + |\mathcal{H}f(x)| + |\mathcal{H}f_1(x)|, \tag{3.2}$$

for almost every $x \in J$. What we have achieved is the liberty to choose $x \in J$. Now we use probabilistic reasoning to prove that there is some point where $|\mathcal{H}f(x)|$ and $|\mathcal{H}f_1(x)|$ are bounded.

By the definition of the Hardy-Littlewood maximal function,

$$\frac{1}{|J|} \int_J |\mathcal{H}f(t)|\, dt \le \mathcal{M}(\mathcal{H}f)(a).$$

The probability that a function is greater than three times its mean is less than $1/3$, therefore

$$\mathfrak{m}\{t \in J : |\mathcal{H}f(t)| < 3\mathcal{M}(\mathcal{H}f)(a)\} \ge \frac{2}{3}|J|.$$

For the other term we use the weak inequality for $\mathcal{H}f_1$

$$\mathfrak{m}\{t \in J : |\mathcal{H}f_1(t)| > \alpha\} \le C\frac{\|f_1\|_1}{\alpha} \le 2C\frac{Mf(a)}{\alpha}|J|.$$

Therefore, in the same way we obtain

$$\mathfrak{m}\{t \in J : |\mathcal{H}f_1(t)| \le 6C Mf(a)\} \ge \frac{2}{3}|J|.$$

Now there is some point $x \in J$ such that, simultaneously

$$|\mathcal{H}f(x)| < 3\mathcal{M}(\mathcal{H}f)(a), \quad \text{and} \quad |\mathcal{H}f_1(x)| < 6C Mf(a).$$

Finally, we arrive at

$$|\mathcal{H}_\varepsilon f(a)| \le C Mf(a) + 3\mathcal{M}(\mathcal{H}f)(a) + 6C Mf(a)$$
$$\le A\big(Mf(a) + \mathcal{M}(\mathcal{H}f)(a)\big).$$

Now we can prove the bound on the norms. Since $p > 1$ we have, applying the known results about the Hardy-Littlewood maximal function and the Hilbert transform,

$$\|\mathcal{H}^* f\|_p \le A\|\mathcal{M}(\mathcal{H}f)\|_p + A\|Mf\|_p \le \frac{Cp}{p-1}\|\mathcal{H}f\|_p + \frac{Cp}{p-1}\|f\|_p$$

$$\le C\frac{p^3}{(p-1)^2}\|f\|_p + \frac{Cp}{p-1}\|f\|_p.$$

For $p > 2$ it follows that

$$\|\mathcal{H}^* f\|_p \le Cp\|f\|_p.$$

$\square$

The inequality that we obtain near $p = 1$ is not sharp. We shall need the sharp estimate to obtain the result of Sjölin. Also we shall give the corresponding result for $p = 1$ that is needed in the proof of the almost everywhere pointwise convergence of $\mathcal{H}_\varepsilon f(x)$ to $\mathcal{H}f(x)$ for $f \in \mathcal{L}^1(\mathbf{R})$.

The problem at $p = 1$ is in Cotlar's Inequality. $\mathcal{M}(\mathcal{H}f)$ may be infinity at every point if $\mathcal{H}f$ is not in $\mathcal{L}^1(\mathbf{R})$. Hence we modify this inequality taking $\mathcal{M}(|\mathcal{H}f|^{1/2})$ instead of $\mathcal{M}(\mathcal{H}f)$.

Theorem 3.8 (Modified Cotlar's Inequality) *Let $f \in \mathcal{L}^1(\mathbf{R})$, then*

$$\mathcal{H}^* f(x) \le A(\{\mathcal{M}(|\mathcal{H}f|^{1/2})(x)\}^2 + \mathcal{M}f(x)).$$

Proof. The proof is the same as that of the inequality of Cotlar, until we obtain inequality (3.2). By definition of the maximal Hardy-Littlewood maximal function,

$$\frac{1}{|J|} \int_J |\mathcal{H}f|^{1/2} \, dm \le \mathcal{M}(|\mathcal{H}f|^{1/2})(a).$$

Therefore

$$\mathfrak{m}\{t \in J : |\mathcal{H}f(t)| < \{3\mathcal{M}(|\mathcal{H}f|^{1/2})(a)\}^2\} \ge \frac{2}{3}\mathfrak{m}(J).$$

The rest of the proof is the same as before. We obtain that there is some point $x \in J$ where simultaneously we have

$$|\mathcal{H}f(x)|^{1/2} < 3\mathcal{M}(|\mathcal{H}f|^{1/2})(a), \text{ and } |\mathcal{H}f_1(x)| \le 6C\mathcal{M}f(a).$$

$\square$

Now we can prove that $\mathcal{H}^* f$ is in weak $\mathcal{L}^1$ when $f \in \mathcal{L}^1(\mathbf{R})$. In fact if $T: \mathcal{L}^1(\mathbf{R}) \to \mathcal{L}_{1,\infty}(\mathbf{R})$, $\mathcal{M}(Tf)$ is not, in general, in $\mathcal{L}_{1,\infty}(\mathbf{R})$, but as Kolmogorov observed (see next proposition), $\mathcal{M}(|Tf|^{1/2})$ is in $\mathcal{L}_{2,\infty}(\mathbf{R})$, and $\|\mathcal{M}(|Tf|^{1/2})\|_{2,\infty} \le \sqrt{2c_1\|T\|\|f\|_1}$. Given the modified Cotlar's inequality, we can prove that $\mathcal{H}^* f$ is in $\mathcal{L}_{1,\infty}(\mathbf{R})$ for every $f \in \mathcal{L}^1(\mathbf{R})$. In fact, $\mathcal{M}(|\mathcal{H}f|^{1/2}) \in \mathcal{L}_{2,\infty}(\mathbf{R})$ gives us

$$\mathfrak{m}\{t \in \mathbf{R} : \{\mathcal{M}(|\mathcal{H}f|^{1/2})\}^2 > \alpha\} \le C\frac{\|f\|_1}{\alpha}.$$

Proposition 3.9 (Kolmogorov) *Let T be an operator such that for every $f \in \mathcal{L}^1(\mathbf{R})$, we have $\|Tf\|_{1,\infty} \le \|T\|\|f\|_1$. Then for every $\alpha > 0$, and $f \in \mathcal{L}^1(\mathbf{R})$*

$$\mathfrak{m}\{t \in \mathbf{R} : \{\mathcal{M}(|Tf|^{1/2})\}^2 > \alpha\} \le c\|T\|\frac{\|f\|_1}{\alpha}.$$

Proof. For every measurable function $g: \mathbf{R} \to \mathbf{C}$ we have

$$g = g[|g| \le \alpha] + g[|g| > \alpha].$$

Therefore $\{\mathcal{M}g > 2\alpha\} \subset \{\mathcal{M}(g[|g| > \alpha]) > \alpha\}$ and by the lemma of Hardy and Littlewood

$$\mathfrak{m}\{\mathcal{M}g > 2\alpha\} \le \frac{c_1}{\alpha} \int_{|g|>\alpha} |g| \, dm.$$

We apply this inequality to our function

$$\mathfrak{m}\{t \in \mathbf{R} : \{\mathcal{M}(|Tf|^{1/2})\}^2 > 4\alpha\} \leq \frac{c_1}{\sqrt{\alpha}} \int_{|Tf|^{1/2} > \sqrt{\alpha}} |Tf|^{1/2} \, dm.$$

Now we follow the traditional path

$$\frac{c_1}{\sqrt{\alpha}} \int_{|Tf|^{1/2} > \sqrt{\alpha}} |Tf|^{1/2} \, dm = \frac{c_1}{2\sqrt{\alpha}} \int_{\alpha}^{+\infty} t^{-1/2} \mathfrak{m}\{|Tf| > t\} \, dt,$$

and now by the hypothesis on T, we get   .

$$\mathfrak{m}\{t \in \mathbf{R} : \{\mathcal{M}(|Tf|^{1/2})\}^2 > 4\alpha\} \leq \frac{c_1\|T\|\|f\|_1}{2\sqrt{\alpha}} \int_{\alpha}^{+\infty} t^{-3/2} \, dt$$

$$= \frac{c_1\|T\|\|f\|_1}{\alpha}.$$

$\square$

Theorem 3.10 *For every $f \in \mathcal{L}^1(\mathbf{R})$ and $\alpha > 0$*

$$\mathfrak{m}\{x \in \mathbf{R} : \mathcal{H}^* f(x) > \alpha\} \leq C \frac{\|f\|_1}{\alpha},$$

where C is an absolute constant.
  Therefore for every $f \in \mathcal{L}^p(\mathbf{R})$

$$\|\mathcal{H}^* f\|_p \leq B \frac{p^2}{p-1} \|f\|_p, \qquad 1 < p < +\infty. \tag{3.3}$$

Proof. By the modified Cotlar's Inequality

$$\{\mathcal{H}^* f > 2\alpha\} \subset \{\mathcal{M}(|\mathcal{H}f|^{1/2}) > \sqrt{\alpha/A}\} \cup \{\mathcal{M}f > \alpha/A\}.$$

Hence, the previous theorem, applied to $\mathcal{H}$ gives us the weak inequality.

Now we know that there is a constant C such that $\|\mathcal{H}f\|_4 \leq C\|f\|_4$. We apply Marcinkiewicz's Theorem to get

$$\|\mathcal{H}^* f\|_p \leq \frac{C}{p-1} \|f\|_p, \qquad 1 < p < 2.$$

Recall that we have proved

$$\|\mathcal{H}^* f\|_p \leq Bp\|f\|_p, \qquad 2 \leq p < +\infty.$$

These two inequalities prove (3.3). $\square$

Finally we arrive at

Theorem 3.11 (Pointwise convergence) *For every $f \in \mathcal{L}^p(\mathbf{R})$, where $1 \le p < +\infty$,*

$$\mathcal{H}f(x) = \lim_{\varepsilon \to 0^+} \mathcal{H}_\varepsilon f(x), \qquad a.\ e.$$

Proof. The proof is almost the same as that of the differentiation theorem. Put

$$\Omega f(x) = \left| \limsup_{\varepsilon \to 0^+} \mathcal{H}_\varepsilon f(x) - \liminf_{\varepsilon \to 0^+} \mathcal{H}_\varepsilon f(x) \right|.$$

What we have to prove is that $\Omega f(x) = 0$ a. e.

For every smooth function of compact support φ it is easily shown that $\Omega\varphi(x) = 0$ for all $x \in \mathbf{R}$. We also know that $\Omega f(x) \le 2\mathcal{H}^* f(x)$ for all $x \in \mathbf{R}$. These two facts combine in the following way

$$\mathfrak{m}\{\Omega f > \alpha\} = \mathfrak{m}\{\Omega(f - \varphi) > \alpha\} \le \mathfrak{m}\{\mathcal{H}^*(f - \varphi) > \alpha/2\}$$

Therefore by the results we have proved about $\mathcal{H}^*$, it follows that

$$\mathfrak{m}\{\Omega f > \alpha\} \le \left(\frac{C\|f - \varphi\|_p}{\alpha} \right)^p.$$

By the density of the smooth functions on $\mathcal{L}^p(\mathbf{R})$, it follows that, for every $\alpha > 0$, $\mathfrak{m}\{\Omega f > \alpha\} = 0$. Therefore $\Omega f(x) = 0$ a. e. as we wanted to prove. $\square$

Part Two

The Carleson–Hunt
Theorem

In this part we study the proof of the Carleson–Hunt Theorem (theorem 12.8 p.172]. To prove the convergence almost everywhere of a Fourier series we consider the maximal operator $S^*(f, x) = \sup_n |S_n(f, x)|$. Instead of this we prefer to consider the **Carleson maximal operator**. This is defined by replacing the Dirichlet kernel by e^{int}/t. In equation (2.3) we have seen that this is almost the same. The Carleson maximal operator is defined as

$$C^* f(x) = \sup_{n \in \mathbf{Z}} \left| \text{p.v.} \int_{-\pi}^{\pi} \frac{e^{in(x-t)}}{x - t} f(t) \, dt \right|.$$

This step follows the observation of Luzin.

We also want to prove that $C^* f(x)$ is in $\mathcal{L}^p(-\pi/2, \pi/2)$ when $f \in \mathcal{L}^p(-\pi, \pi)$. By the interpolation theorems we must bound the measure of the set where $C^* f(x) > y$.

Therefore we are interested in bounding the **Carleson integrals**

$$\text{p.v.} \int_{-\pi}^{\pi} \frac{e^{in(x-t)}}{x - t} f(t) \, dt.$$

The main idea of the proof is to decompose the interval I into subintervals, one of them, $I(x)$, containing x ($x \in I(x)/2$). Then we have

$$\text{p.v.} \int_{I} \frac{e^{i\lambda(x-t)}}{x - t} f(t) \, dt = \text{p.v.} \int_{I(x)} \frac{e^{i\lambda(x-t)}}{x - t} f(t) \, dt + \sum_{J} \int_{J} \frac{e^{i\lambda(x-t)}}{x - t} f(t) \, dt.$$

Clearly, the first term gives the main contribution. It is almost a new Carleson integral, the only difference being that, in general, the number of cycles will not be an integer. We shall see that we can replace it by a Carleson integral.

The other terms can be conveniently bounded. To this end we write

$$\int_{J} \frac{e^{i\lambda(x-t)}}{x - t} f(t) \, dt = \int_{J} \frac{e^{i\lambda(x-t)} f(t) - M}{x - t} \, dt + \int_{J} \frac{M}{x - t} \, dt,$$

where M is the mean value of $e^{i\lambda(x-t)} f(t)$ on J.

Now we can benefit from the fact that the integral of $e^{i\lambda(x-t)} f(t) - M$ in J is 0, and put

$$\int_{J} \frac{e^{i\lambda(x-t)} f(t) - M}{x - t} \, dt = \int_{J} \{ e^{i\lambda(x-t)} f(t) - M \} \left(\frac{1}{x - t} - \frac{1}{x - t_J} \right) dt,$$

where t_J is the center of J.

All these terms can be bounded by weak local norms

$$\|f\|_{(n,J)} = \sum_{j \in \mathbf{Z}} \frac{c}{1+j^2} \left| \frac{1}{|J|} \int_J f(t) \exp\left(-2\pi i \left(n + \frac{j}{3}\right) \frac{t}{|J|}\right) dt \right|.$$

Hence, given $\alpha = (n, J)$, $\|f\|_\alpha$ is a mean value of absolute values of (generalized) Fourier coefficients of f.

Observe now that the first term is of the type with which we started, only that the new integral is more simple because it corresponds to fewer cycles, but a non integer number of cycles. We can change to a new integral that has an integer number of cycles, and in such a way that the difference is again bounded by the local norm.

We iterate this process until we obtain a Carleson integral with 0 cycles, that is, we arrive to a Hilbert transform that is easily estimated.

The remainders that appear in these steps are bounded except on an **exceptional set**. The larger we allow the bound, the smaller can this exceptional set be made.

To bound the number of steps we start with $n \le 2^N$. In this way we arrive naturally to the log log n result. That is because for every interval that appears in the process we must add a fraction to the exceptional set. In the process we consider every dyadic interval of length $\le 2\pi/2^N$. This gives a contribution of length proportional to N. We must compensate with log N in the bound allowed.

Chapter four contains the **basic step**. This is a precise formulation of how we choose the partition of the interval of a given Carleson integral, and the bounds we obtain. It also contains the theorem of **change of frequency** that gives a good bound of the difference between two Carleson integrals in the same interval but with different frequency. This theorem plays an important role in the second part of the proof. Chapter five gives the bounds of the terms that appear in the basic step. And finally chapter six defines the exceptional set and combines the previous bounds to obtain the result $S_n(f, x) = o(\log \log n)$ a.e. for every $f \in \mathcal{L}^2[-\pi, \pi]$.

Then we start the proof of the Carleson-Hunt theorem. This is a refined version of the previous theorem. An analysis of the previous proof shows the reason for the log log n term: we have used in the proof all the pairs $\alpha = (n, I)$. But if we think about the process of selection of the central interval $I(x)$ we notice that, in it, the local norm must be relatively great. Therefore we need to select a set S of **allowed pairs** and assure that we only use them in the inductive steps.

If we start with a Carleson integral (maybe one for an allowed pair), and carry out the procedure of chapter four, we reach a Carleson integral that in general does not correspond to an allowed pair. Therefore we must change to another one, but with a controllable error.

In the way to the definition of the allowed pairs Carleson gives an **analysis of the function** f that is very clever. This analysis can be regarded as that of writing the score from a piece of music. Given a level $b_j y^{p/2}$, (the intensity of the least intense note), we start from the four equal intervals into which the interval I is divided. In every one we obtain the Fourier series of f, and retain only those terms that are greater that the chosen level. This forms the polynomial P_j. Then every one of these intervals is divided into two parts. In every one of these parts we obtain the Fourier series of $f - P_j$, and proceed in the same way.

This procedure gives one a set of pairs Q that we call **notes** of f. This will be the starting point for the definition of the allowed pairs

In chapters 4, 5 and 6 we will apply the principal idea of the proof in a straightforward way. We will obtain that for every $f \in \mathcal{L}^2$ the partial sums $S_n(f, x) = o(\log \log n)$. This reasoning is relatively easy to follow. The kernel of the proof is in chapters 7 to 10. In these chapters the previous proof is refined to remove the $\log \log n$ term.

4. The Basic Step

4.1 Introduction

The proof of the Carleson Theorem is based on a new method of estimating partial sums of Fourier series. We replace the Dirichlet integral by the singular integral

$$\mathcal{C}_{(n,I)} f(x) = \text{p.v.} \int_{-\pi}^{\pi} \frac{e^{in(x-t)}}{x-t} f(t)\, dt.$$

The new method consists of applying repeatedly a basic step that we are going to analyze in this chapter.

Given a partition Π of the interval $I = [-\pi, \pi]$, we decompose the integral as

$$
\mathcal{C}_{(n,I)} f(x) = \text{p.v.} \int_{I(x)} \frac{e^{in(x-t)}}{x-t} f(t)\, dt + \sum_{J \in \Pi, J \not\subset I(x)} \int_J \frac{E_\Pi f(t)}{x-t}\, dt
$$
$$
+ \sum_{J \in \Pi, J \not\subset I(x)} \int_J \frac{e^{in(x-t)} f(t) - E_\Pi f(t)}{x-t}\, dt,
$$

(4.1)

where $I(x)$ is an interval containing x, that is a union of some members of Π, and $E_\Pi f$ is the conditional expectation of $e^{in(x-t)} f(t)$ with respect to Π.

The principal point of the basic step is the careful choice of $I(x)$ and Π so that we have good control of the last two sums. The first term is an integral of the same type that we can treat in a similar way.

First we define local norms so that we can give adequate bounds for the last two terms. Later we will show the precise bounds and we will give the form of the basic step in theorem 4.13.

4.2 Carleson maximal operator

Let $I \subset \mathbf{R}$ be a bounded interval and $n \in \mathbf{Z}$. For every $f \in \mathcal{L}^1(I)$, we consider the singular integral

$$\text{p.v.} \int_I \frac{e^{2\pi i n(x-t)/|I|}}{x-t} f(t)\, dt.$$

(4.2)

If x is contained in the interval $I/2$ with the same center as I and half length we call (4.2) a **Carleson integral**.

Here we have an arbitrary selection: $x \in I/2$. Every condition $x \in \theta I$, with $\theta \in (0, 1)$ would be adequate. If we allow x to be near the extremes of I, the simplest Carleson integral would be unbounded. For example if we want to bound

$$\sup\left|\text{p.v.} \int_{-a}^{b} \frac{dt}{t}\right| = \sup_{a,b>0}\left|\log\frac{b}{a}\right|$$

we need $M^{-1} < b/a < M$. This is the condition $0 \in \theta(a, b)$ with $\theta = (M - 1)/(M + 1) \in (0, 1)$. Although this selection is arbitrary, many details of the proof shall depend on the selection made here.

The set of pairs $\mathcal{P} = \{(n, I) : n \in \mathbf{Z}, I$ a bounded interval of $\mathbf{R}\}$ will assume a very central role in the following. We use greek letters to denote the elements of $\mathcal{P}$. Given a pair $\alpha = (n, I)$ we denote I by $I(\alpha)$ and n by $n(\alpha)$. Also we call $|I| = |I(\alpha)|$ the **length** of α, and write it as $|\alpha|$. Then for every $\alpha \in \mathcal{P}$ and $x \in I(\alpha)/2$ we put

$$\mathcal{C}_\alpha f(x) = \text{p.v.} \int_{I(\alpha)} \frac{e^{2\pi i n(\alpha)(x-t)/|\alpha|}}{x - t} f(t)\, dt.$$

Given $f \in \mathcal{L}^1(I)$ and $\alpha \in \mathcal{P}$ with $I(\alpha) = I$, $\mathcal{C}_\alpha f(x)$ is defined for almost every $x \in I(\alpha)/2$. We shall study the maximal Carleson operator

$$C_I^* f(x) = \sup_{I(\alpha)=I, n(\alpha)\in\mathbf{Z}} |\mathcal{C}_\alpha f(x)|.$$

This is a measurable function with values in $[0, +\infty]$ (since it is a supreme of a countable number of measurable functions). We shall prove that it is a bounded operator from $\mathcal{L}^p(I)$ to $\mathcal{L}^p(I/2)$ for every $1 < p < +\infty$.

The following proposition gives us some practice with Carleson integrals, and it will be needed in the proof of Carleson's Theorem:

Lemma 4.1 *There exists an absolute constant $A > 0$ such that if $x \in I(\gamma)/2$, and for every $\lambda \in \mathbf{R}$,*

$$|\mathcal{C}_\gamma(e^{i\lambda t})(x)| \leq A.$$

Proof. Write $I = I(\gamma)$. By definition if $\lambda(\gamma) = 2\pi n(\gamma)/|\gamma|$,

$$\mathcal{C}_\gamma(e^{i\lambda t})(x) = \text{p.v.} \int_I \frac{e^{i\lambda(\gamma)(x-t)}}{x - t} e^{i\lambda t}\, dt.$$

Hence, if $\omega = \lambda(\gamma) - \lambda$,

$$|\mathcal{C}_\gamma(e^{i\lambda t})(x)| = \left|\text{p.v.} \int_I \frac{e^{i\omega(x-t)}}{x - t}\, dt\right|.$$

We change variables, letting $u = x - t$ and see that

$$|C_\gamma(e^{i\lambda t})(x)| = \left|\text{p.v.} \int_{-a}^{b} \frac{e^{i\omega u}}{u} \, du\right|.$$

Since $x \in I(\gamma)/2$, we have $0 \in J/2$ if $J = [-a, b]$. This condition is equivalent to $1/3 < b/a < 3$. Therefore

$$|C_\gamma(e^{i\lambda t})(x)| = \left|\text{p.v.} \int_{-a}^{a} \frac{e^{i\omega u}}{u} \, du + \int_{a}^{b} \frac{e^{i\omega u}}{u} \, du\right|$$

$$\leq \left|\int_{-a}^{a} \frac{\sin \omega u}{u} \, du\right| + \log 3.$$

Furthermore $\left|\int_{-x}^{x}(\sin t/t)\, dt\right|$ is bounded, thus proving our lemma. □

4.3 Local norms

The basic procedure is the reduction of one Carleson integral to another, simpler, Carleson integral. In order to bound the difference between them we shall need norms associated to pairs $\alpha \in \mathcal{P}$.

First we associate a function e_α to every pair $\alpha \in \mathcal{P}$. e_α is a function in $\mathcal{L}^2(\mathbf{R})$ supported by the interval $I(\alpha)$.

$$e_\alpha(x) = \exp\left(2\pi i \frac{n(\alpha)}{|\alpha|} x\right) \chi_{I(\alpha)}(x) = e^{i\lambda(\alpha)x} \chi_{I(\alpha)}(x).$$

The function e_α is a **localized wave train**. It is localized in the **time interval** $I(\alpha)$, and has **angular frequency** $\lambda(\alpha) = 2\pi n(\alpha)/|I(\alpha)|$. The number $|\alpha| = |I(\alpha)|$ is the **duration** and $n(\alpha)$ the total **number of cycles** in the wave train.

The functions e_α with $I(\alpha) = I$ fixed, form an orthonormal system and every function f supported by I can be developed in a series of these functions, convergent in $\mathcal{L}^2(\mathbf{R})$.

$$f = \sum_{I(\alpha)=I} \frac{\langle f, e_\alpha \rangle}{|\alpha|} e_\alpha.$$

Observe that

$$\frac{\langle f, e_\alpha \rangle}{|\alpha|} = \frac{1}{|\alpha|} \int_{I(\alpha)} f(t) \exp\left(-2\pi i \frac{n(\alpha)}{|\alpha|} t\right) dt.$$

We define the local norm $\|f\|_\alpha$ as

$$\|f\|_\alpha = \sum_{j \in \mathbf{Z}} \frac{c}{1 + j^2} \left|\frac{1}{|I|} \int_{I} f(t) \exp\left(-2\pi i \left(n(\alpha) + \frac{j}{3}\right) \frac{t}{|I|}\right) dt\right|.$$

Here c is chosen so that $\sum_{j \in \mathbf{Z}} c/(1+j^2) = 1$. Hence $\|f\|_\alpha$ is a mean value of absolute values of (generalized) Fourier coefficients of f.

One motivation for this definition is that with this norm we can control the integrals $\int_I f(t)\varphi(t)\,dt$ when the function φ is twice continuously differentiable on I. We shall show this in theorem 4.3 below.

Proposition 4.2 *Let $\varphi \in \mathcal{C}^2[a,b]$, $\delta = b - a$. For every $x \in [a,b]$ we have*

$$\varphi(x) = \sum_{n \in \mathbf{Z}} c_n \exp\left(2\pi i \frac{n}{3}\frac{x}{\delta}\right),$$

where the coefficient c_n satisfies $(1+n^2)|c_n| \le A(\|\varphi\|_\infty + \delta^2\|\varphi''\|_\infty)$, for every $n \in \mathbf{Z}$.

Proof. First assume that $\delta = 1$. We can extends φ to a twice continuously differentiable function $\tilde\varphi$, of period 3, defined on $\mathbf{R}$. We can assume that $\|\tilde\varphi\|_\infty$, $\|\tilde\varphi'\|_\infty$ and $\|\tilde\varphi''\|_\infty$ are bounded by $\|\varphi\|_\infty + \|\varphi''\|_\infty$.

The Fourier series for $\tilde\varphi$ is

$$\tilde\varphi(x) = \sum_n c_n \exp\left(2\pi i \frac{n}{3} x\right),$$

where

$$c_n = \frac{1}{3}\int_0^3 \tilde\varphi(t)\exp\left(-2\pi i \frac{n}{3} t\right) dt = \int_0^1 \tilde\varphi(3t) e^{-2\pi i n t}\,dt.$$

Now, integration by parts leads, for $n \ne 0$ to

$$c_n = \frac{9}{(2\pi i n)^2}\int_0^1 \tilde\varphi''(3t)e^{-2\pi i n t}\,dt,$$

so that for some absolute constant A and every $n \in \mathbf{Z}$

$$(1+n^2)|c_n| \le A(\|\varphi\|_\infty + \|\varphi''\|_\infty).$$

In the general case a change of scale leads to the inequality

$$(1+n^2)|c_n| \le A(\|\varphi\|_\infty + \delta^2\|\varphi''\|_\infty).$$

$\square$

Theorem 4.3 *Let $f \in \mathcal{L}^1(I)$, $\varphi \in \mathcal{C}^2(I)$ and $\alpha = (n, I) \in \mathcal{P}$. For some absolute constant B we have*

$$\left|\frac{1}{|\alpha|}\int_I e^{2\pi i n(x-t)/|\alpha|} f(t)\varphi(t)\,dt\right| \le B(\|\varphi\|_\infty + |\alpha|^2\|\varphi''\|_\infty)\,\|f\|_\alpha.$$

Proof. We apply proposition 4.2 to φ and obtain

$$\varphi(x) = \sum_{j \in \mathbf{Z}} c_j \exp\left(2\pi i \frac{j}{3} \frac{x}{|\alpha|}\right),$$

where $(1 + j^2)|c_j| \le A(\|\varphi\|_\infty + |\alpha|^2 \|\varphi''\|_\infty)$. Hence we have

$$\left| \frac{1}{|\alpha|} \int_I e^{2\pi i n(x-t)/|\alpha|} f(t)\varphi(t)\, dt \right|$$

$$\le \sum_{j \in \mathbf{Z}} |c_j| \left| \frac{1}{|\alpha|} \int_I f(t) \exp\left(-2\pi i \left(n - \frac{j}{3}\right) \frac{t}{|\alpha|}\right) dt \right|$$

$$\le B(\|\varphi\|_\infty + |\alpha|^2 \|\varphi''\|_\infty) \|f\|_\alpha.$$

$\square$

We shall need a bound of $\|f\|_\alpha$ when f is an exponential, and will give it here:

Proposition 4.4 *There exists an absolute constant C such that for every $\omega \in \mathbf{R}$ and $\alpha \in \mathcal{P}$ we have*

$$\|e^{2\pi i \omega x}\|_\alpha \le 1, \qquad \text{and} \qquad \|e^{2\pi i \omega x}\|_\alpha \le \frac{C}{|\lfloor \omega|\alpha| \rfloor - n(\alpha)|}.$$

Proof. Since $|e^{2\pi i \omega x}| = 1$ and $\|f\|_\alpha$ is a mean value of integrals,

$$\left| \frac{1}{|\alpha|} \int_{I(\alpha)} f(t) \exp\left(-2\pi i \left(n(\alpha) + \frac{j}{3}\right) \frac{t}{|\alpha|}\right) dt \right|, \tag{4.3}$$

the first inequality is obvious.

For the second inequality we calculate (4.3), and obtain

$$\frac{|e^{2\pi i (A - j/3)} - 1|}{2\pi |A - j/3|},$$

where $A = \omega|\alpha| - n(\alpha)$. Hence we have

$$\|e^{2\pi i \omega x}\|_\alpha \le \sum_{j \in \mathbf{Z}} \frac{c}{1 + j^2} \frac{|e^{2\pi i (A - j/3)} - 1|}{2\pi |A - j/3|}$$

$$\le \frac{1}{2\pi} \sum_{|j|/3 \le |A|/2} \frac{c}{1 + j^2} \frac{4}{|A|} + \sum_{|j|/3 > |A|/2} \frac{c}{1 + j^2} \le \frac{2}{\pi |A|} + \frac{M}{|A|}.$$

And, therefore

$$\|e^{2\pi i \omega x}\|_\alpha \le \frac{(2/\pi) + M}{|A|}.$$

For technical reasons it will be convenient to replace $A = \omega|\alpha| - n(\alpha)$ by $\lfloor \omega|\alpha| \rfloor - n(\alpha)$. Since $\|e^{2\pi i \omega x}\|_\alpha \le 1$ always, this presents no problem. $\square$

Proposition 4.5 *There exists an absolute constant $B > 0$ such that, for every interval K and $\omega \in \mathbf{R}$, there exists $k \in \mathbf{N}$ such that for $\kappa = (k, K)$, we have*

$$\|e^{2\pi i \omega x}\|_\kappa \geq B.$$

We can choose $k = \lfloor |\omega| \cdot |K| \rfloor$.

Proof. As in the proof of proposition 4.4 we obtain

$$\|e^{2\pi i \omega x}\|_\kappa = \sum_{j \in \mathbf{Z}} \frac{c}{1 + j^2} \frac{|e^{2\pi i(A - j/3)} - 1|}{2\pi |A - j/3|} \geq \frac{c}{1 + j_0^2} \frac{|e^{2\pi i(A - j_0/3)} - 1|}{2\pi |A - j_0/3|}.$$

One of every three consecutive j satisfies $|e^{2\pi i(A - j/3)} - 1| > 1$. Hence we can choose j_0 such that $|A - j_0/3| < 1$, and $|e^{2\pi i(A - j_0/3)} - 1| > 1$. For such j_0 we have

$$\|e^{2\pi i \omega x}\|_\kappa \geq \frac{C}{1 + 9(1 + |A|)^2}.$$

This is greater than B if we choose $|A| < 1$. Since $A = |\omega| \cdot |K| - n(\kappa)$, this goal is achieved taking $k = n(\kappa) = \lfloor |\omega| \cdot |K| \rfloor$.   □

Note that for $\alpha = (n, I)$ and $f \in \mathcal{L}^p(I)$ we have

$$\|f\|_\alpha \leq \|f\|_{\mathcal{L}^p(I)}, \tag{4.4}$$

Here and elsewhere if $f \in \mathcal{L}^p(I)$ we put

$$\|f\|_{\mathcal{L}^p(I)} = \left(\frac{1}{|I|} \int_I |f|^p \, dm \right)^{1/p}, \quad \text{and} \quad \|f\|_p = \left(\int_I |f|^p \, dm \right)^{1/p}.$$

Hence $\|f\|_{\mathcal{L}^p(I)}$ always denotes the $\mathcal{L}^p(I)$ norm with respect to normalized Lebesgue measure.

4.4 Dyadic partition

Given an interval $I = [a, b]$, the central point c of I divides this interval in two intervals of half length that we denote by $I_0 = [a, c]$ and $I_1 = [c, b]$.

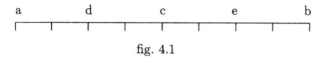

fig. 4.1

Analogously given $I_1 = [c, b]$ we obtain the interval $(I_1)_0 = I_{10} = [c, e]$ and $I_{11} = [e, b]$ where e is the central point of the interval I_1.

The same process defines the intervals I_u for every word $u \in \{0,1\}^*$. We call these intervals *dyadic intervals* generated from I.

Every dyadic interval I_u has two *sons*, I_{u0} and I_{u1}. Every dyadic interval has a *brother*. For example, the brother of I_{00101} is I_{00100}. But, in general, every dyadic interval has two *contiguous* intervals. We understand by contiguous an interval of the same length and with a unique point in common. For example it is easy to see that the contiguous intervals of I_{00101} are its brother I_{00100} and I_{00110}. We also speak of the *grandsons* of I. They are the four intervals, I_{00}, I_{01}, I_{10}, and I_{11}.

We are dealing with the operator $C_I^*: \mathcal{L}^p(I) \to \mathcal{L}^p(I/2)$. I denotes always this interval and we speak of *dyadic* intervals as those intervals that can be written as I_u with $u \in \{0,1\}^*$. Also it must be noticed that I is an arbitrary interval, so we can apply every concept defined on I to every other interval. For example, we can speak of dyadic intervals with respect to J.

Smoothing intervals. In general the union of two contiguous dyadic intervals is not a dyadic interval. Such an interval we shall call a *smoothing interval*. They play a prominent role in the proof. All dyadic intervals are smoothing intervals, since they can be written as the union of its two sons. But there are smoothing intervals that are not dyadic. For example the interval $[d,e] = I_{01} \cup I_{10}$ is a smoothing interval, but it is not dyadic.

Given an interval I we shall denote by $I/2$ its middle half, that is $I/2 = I_{01} \cup I_{10}$.

Given the interval I, we denote by $\mathcal{P}_I$ the set of pairs (n, J) where $n \in \mathbf{Z}$ and J is a smoothing interval with respect to I.

Dyadic Points. The extremes of all dyadic intervals (with respect to I) are called dyadic points and the set of all dyadic points will be denoted by D. D is a countable set and so it is of measure 0.

Proposition 4.6 *Given I and $x \in I/2$ that is not a dyadic point, for every $n = 0, 1, 2, \ldots$ there is only one smoothing interval I_n of length $|I|/2^n$ such that $x \in I_n/2$. We also have $I = I_0 \supset I_1 \supset I_2 \supset \ldots$.*

Proof. It is clear that for $n = 0$ there is only one smoothing interval I. Now assume that there is only one smoothing interval $J = I_n = [a,b]$ with $x \in I_n/2$. Then, $x \in (d,e)$ (refer to the figure 1.1). Since $x \notin D$, there is one and only one interval J_{010}, J_{011}, J_{100}, or J_{101} that contains x. In every case we check that there is only one smoothing interval with the required conditions, I_{n+1} will be respectively $[a,c]$, $[d,e]$, $[d,e]$, or $[c,b]$. We observe that in every case $I_{n+1} \subset I_n$. $\square$

Choosing $I(x)$. We shall consider partitions Π of some smoothing interval J where every member of Π is a dyadic interval generated from I, of length $\leq |J|/4$. We always consider closed intervals, and when we speak of partitions we don't take into consideration the extremes of these intervals. We also speak of *disjoint* intervals to mean those whose interiors are disjoint.

We assume a Carleson integral $C_\alpha f(x)$ and a dyadic partition Π of $I = I(\alpha)$, where every $J \in \Pi$ has length $|J| \leq |I|/4$, to be given. $I(x)$ will be an interval, a union of some members of Π, such that $x \in I(x)/2$, so that the first term in decomposition (4.1) will be *almost* a Carleson integral. We must also choose $I(x)$ in order to obtain a good bound for the other members of decomposition (4.1). For example

$$\int_J \frac{E_\Pi f(t)}{x - t}\, dt.$$

Here $x \notin J \in \Pi$. Therefore we shall choose $I(x)$ so that $|x - t|$ is of the same order as $|J|$, for every $t \in J$.

The next proposition guarantees that all these conditions can be attained.

Proposition 4.7 *Let $x \in I/2$ and let Π be a dyadic partition of the smoothing interval I, with intervals of length $\leq |I|/4$. Then there exists a smoothing interval $I(x)$ such that:*

(a) $x \in I(x)/2$.

(b) $|I(x)| \leq |I|/2$.

(c) $I(x)$ is the union of some intervals of Π.

(d) Some of the two sons of $I(x)$ is a member of Π.

(e) For every $J \in \Pi$ such that $J \not\subset I(x)$ we have $d(x, J) \geq |J|/2$.

(f) Each smoothing interval J with $I(x) \subset J \subset I$ and $x \in J/2$ is a union of intervals of the partition Π.

Proof. We consider the smoothing intervals $J \cup J'$, union of an interval $J \in \Pi$ and a contiguous interval J', such that $x \in (J \cup J')/2$. For example if $x \in J_0 \in \Pi$, we can choose J_0' contiguous to J_0 and such that $J_0 \cup J_0'$ satisfies these conditions. Now let $I(x)$ be such an interval $J \cup J'$ of maximum length. Then (a) and (d) are satisfied by construction. (b) follows from the hypothesis that every $J \in \Pi$ has length $\leq |I|/4$.

(c) Let $I(x) = J \cup J'$ with $J \in \Pi$. J and J' are dyadic intervals.

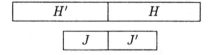

fig. 4.2

Two dyadic intervals have intersection of measure zero or one is contained in the other. Hence if J' is not the union of intervals of Π, there must be an element $H \in \Pi$ such that $J' \subset H$. Since H and J are members of Π they are essentially disjoint. Let H' be the dyadic interval of the same length as H that contains J. As J and J' are contiguous, H and H' are also contiguous.

Then $K = H \cup H'$ satisfies $x \in K/2$, and $|K| > |I(x)|$. This contradicts the definition of $I(x)$.

(e) Let $J \in \Pi$ be such that $J \cap I(x)$ is of measure zero. If $d(x, J) < |J|/2$, there exists J' contiguous to J and containing x. Hence $x \in (J \cup J')/2$. By definition this implies $I(x) \supset J \cup J'$, which contradicts the hypothesis that $J \not\subset I(x)$.

(f) Let J be such a smoothing interval and let K be a son of J. Two dyadic intervals are disjoint or one is contained in the other. Since K is dyadic it follows that K is the union of the intervals $L \in \Pi$ such that $L \subset K$ or there is $L \in \Pi$ with $L \supsetneq K$. The case $L \supsetneq K$ is impossible, because if $L \supsetneq K$, since $x \in J/2$ we have

$$d(x, L) \le d(x, K) \le |K|/2 \le \frac{1}{4}|L|.$$

Therefore there is L' contiguous to L and such that $x \in (L \cup L')/2$. Also $|L \cup L'| > 2|K| \ge |I(x)|$, which contradicts the definition of $I(x)$. $\qquad \square$

The condition (e) says that every interval $J \in \Pi$ such that $J \not\subset I(x)$ is contained in a security interval J' of length $2|J|$ such that x is not contained in J'. This will play a role in the bound (4.15).

4.5 Some definitions

The remainder terms in (4.1) will be bounded using two functions. The first is a modification of the maximal Hilbert transform.

Definition 4.8 Given $f \in \mathcal{L}^1(I)$, we define $H_I^* f(x)$, the maximal Hilbert dyadic transform on I, by

$$H_I^* f(x) = \sup_K \left| \text{p.v.} \int_K \frac{f(t)}{x - t}\, dt \right|,$$

where the supremum is over the intervals $K = J \cup J'$, which are the union of two contiguous dyadic intervals such that $x \in K/2$.

The second function will be needed to bound the third term in (4.1).

Definition 4.9 Given a finite partition Π of I by intervals J_k of length δ_k and center t_k we define the function

$$\Delta(\Pi, x) = \sum_k \frac{\delta_k^2}{(x - t_k)^2 + \delta_k^2}.$$

In the future reasoning a pair $\alpha = (n, I)$, (with some other elements that now are of no consequence) will determine a dyadic partition Π_α of $I(\alpha)$. Hence we shall denote by $\Delta_\alpha(x)$ the corresponding function $\Delta(\Pi_\alpha, x)$. We shall also use $H_\alpha^* f(x)$ to denote the maximal Hilbert dyadic transform $H_{I(\alpha)}^* E_\alpha f(x)$, where the function $E_\alpha f(x)$ denotes

$$\mathbb{E}(e^{i\lambda(\alpha)(x-t)} f(t), \Pi_\alpha),$$

the expectation of the function $e^{i\lambda(\alpha)(x-t)} f(t)$ with respect to the partition Π_α. Hence if $z \in J \in \Pi$, $E_\alpha f(z) = |J|^{-1} \int_J e^{i\lambda(\alpha)(x-t)} f(t) \, dt$. (In order not to be pedantic, the notations Δ_α, and $H_\alpha^* f$ do not mention the partition Π_α).

In the sequel we analyze every term of the decomposition (4.1). This will allow us to decide how to choose our partition Π_α.

4.6 Basic decomposition

The basic step in the proof of Carleson's Theorem is a decomposition of a Carleson integral $\mathcal{C}_\alpha f(x)$ into three parts associated with a dyadic partition Π of $I = I(\alpha)$. We assume that every $J \in \Pi$ has a measure $\leq |I|/4$; then Proposition 4.7 gives us an interval $I(x)$ such that $x \in I(x)/2$.

The decomposition is given by

$$\mathcal{C}_\alpha f(x) = \text{p.v.} \int_I \frac{e^{i\lambda(\alpha)(x-t)}}{x-t} f(t) \, dt$$

$$= \text{p.v.} \int_{I(x)} \frac{e^{i\lambda(\alpha)(x-t)}}{x-t} f(t) \, dt$$

$$+ \int_{I \setminus I(x)} \frac{E_\alpha f(t)}{x-t} \, dt + \int_{I \setminus I(x)} \frac{e^{i\lambda(\alpha)(x-t)} f(t) - E_\alpha f(t)}{x-t} \, dt.$$

$$(4.5)$$

We shall transform the first term into a Carleson integral that can be considered simpler than $\mathcal{C}_\alpha f(x)$. The second and third terms can be bounded in terms of the functions $H_\alpha f(x)$ and $\Delta_\alpha(x) = \Delta(\Pi_\alpha, x)$.

In the following sections we are going to obtain the relevant bounds. These are collected together in theorem 4.11, that is a preliminary version of the basic step. Then we choose a partition in order to optimize the bound of theorem 4.11. With the selected partition Π we formulate a new form of the basic step in Theorem 4.13.

In what follows we will try to bound Carleson integrals $\mathcal{C}_\alpha f(x)$. It must be understood that we put $|\mathcal{C}_\alpha f(x)| = +\infty$ if the principal value is not defined.

With this convention it can be noticed that our bounds are also correct in this case.

4.7 The first term

In fact, the first integral in (4.5) is almost a Carleson integral. We need $\beta = (m, I(x)) \in \mathcal{P}$ such that the difference between $\mathcal{C}_\beta f(x)$ and the first integral is small. Now these are

$$\text{p.v.} \int_{I(x)} \frac{e^{i\lambda(\alpha)(x-t)}}{x-t} f(t)\, dt, \qquad \text{p.v.} \int_{I(x)} \frac{e^{i\lambda(\beta)(x-t)}}{x-t} f(t)\, dt.$$

They will be equal if $\lambda(\alpha) = \lambda(\beta)$, that is,

$$2\pi \frac{n}{|I|} = 2\pi \frac{m}{|I(x)|}. \tag{4.6}$$

In general it is not possible to choose $m \in \mathbf{Z}$ such that (4.6) holds. Hence we choose

$$m = \left\lfloor n \frac{|I(x)|}{|I|} \right\rfloor. \tag{4.7}$$

Now that we have chosen a convenient $\beta = (m, I(x)) \in \mathcal{P}$ we must bound the difference.

$$\text{p.v.} \int_{I(x)} \frac{e^{i\lambda(\alpha)(x-t)}}{x-t} f(t)\, dt - \mathcal{C}_\beta f(x).$$

Here, for the first time, we want to change the frequency of a Carleson integral. These changes are governed by

Theorem 4.10 (Change of frequency) *Let α, β be two pairs with $I(\alpha) = I(\beta) = J$ and $x \in J/2$. If $|n(\alpha) - n(\beta)| \leq M$ with $M > 1$, then*

$$|\mathcal{C}_\beta f(x) - \mathcal{C}_\alpha f(x)| \leq BM^3 \|f\|_\alpha, \tag{4.8}$$

where B is some absolute constant.

Remark. It is not necessary that $n(\beta) \in \mathbf{Z}$.

Proof. We apply Theorem 4.3. First observe that

$$|\mathcal{C}_\beta f(x) - \mathcal{C}_\alpha f(x)| = \left| \int_J \frac{e^{i\left(\lambda(\beta) - \lambda(\alpha)\right)(x-t)} - 1}{x-t} e^{i\lambda(\alpha)(x-t)} f(t)\, dt \right|.$$

Let $\lambda(\beta) - \lambda(\alpha) = L$; then

$$|\mathcal{C}_\beta f(x) - \mathcal{C}_\alpha f(x)| = |L|\,|\alpha| \left| \frac{1}{|J|} \int_J \frac{e^{iL(x-t)} - 1}{|L|(x-t)} e^{i\lambda(\alpha)(x-t)} f(t)\,dt \right|.$$

The function $(e^{it} - 1)/t$ defined on $\mathbf{R}$ is in $\mathcal{C}^\infty(\mathbf{R})$ and so it is easy to see that if $\varphi(t) = (e^{iL(x-t)} - 1)/L(x-t)$ then

$$\|\varphi\|_\infty + |\alpha|^2 \|\varphi''\|_\infty \leq C(1 + L^2 |\alpha|^2).$$

Since $|L|\,|\alpha| = 2\pi |n(\beta) - n(\alpha)|$, Theorem 4.3 gives us

$$|\mathcal{C}_\beta f(x) - \mathcal{C}_\alpha f(x)| \leq C|L|\,|\alpha|(1 + |L|^2|\alpha|^2)\|f\|_\alpha \leq BM^3\|f\|_\alpha.$$

$\square$

If we apply this to our case it follows that

$$\left| \text{p.v.} \int_{I(x)} \frac{e^{i\lambda(\alpha)(x-t)}}{x-t} f(t)\,dt - \mathcal{C}_\beta f(x) \right| \leq C\|f\|_\beta. \tag{4.9}$$

4.8 Notation α/β

We have related the first term in (4.5) to a Carleson integral $\mathcal{C}_\beta f(x)$. The process to obtain β from α and $I(x)$ will be used very often. Hence given $\alpha \in \mathcal{P}$ and an interval $J \subset I(\alpha)$ we define $\alpha/J \in \mathcal{P}$ by

$$\alpha/J = (m, J), \qquad \text{where} \qquad m = \left\lfloor n(\alpha)\frac{|J|}{|\alpha|} \right\rfloor.$$

We choose α/J so that $e_{\alpha/J}$ represents more or less the same *musical note* as e_α, but of duration J, as far as this is possible.

Observe that by definition

$$0 \leq \big(\lambda(\alpha) - \lambda(\alpha/J)\big)|J| < 2\pi. \tag{4.10}$$

Also, given α and $\beta \in \mathcal{P}$ such that $I(\beta) \subset I(\alpha)$, we define

$$\alpha/\beta = \alpha/I(\beta).$$

For future reference we notice the following relation:

$$\frac{\lambda(\alpha)}{2\pi}|\beta| = n(\alpha/\beta) + h, \qquad \text{where } 0 \leq h < 1, \tag{4.11}$$

valid whenever α and $\beta \in \mathcal{P}$ are such that $I(\beta) \subset I(\alpha)$.

This follows from the identity

$$\frac{\lambda(\alpha)}{2\pi}|\beta| = n(\alpha)\frac{|\beta|}{|\alpha|}.$$

If we have $I(\alpha) \supset J \supset K$ and all are smoothing intervals, we have

$$\alpha/K = (\alpha/J)/K. \tag{4.12}$$

In fact, we have

$$n(\alpha/K) = \left\lfloor n\frac{|K|}{|\alpha|} \right\rfloor, \quad n((\alpha/J)/K) = \left\lfloor n(\alpha/J)\frac{|K|}{|J|} \right\rfloor = \left\lfloor \left\lfloor n\frac{|J|}{|\alpha|} \right\rfloor \frac{|K|}{|J|} \right\rfloor.$$

All the lengths are of type $|I|/2^s$. Hence all we have to prove is

$$\left\lfloor \frac{\lfloor n/2^k \rfloor}{2^l} \right\rfloor = \left\lfloor \frac{n}{2^{k+l}} \right\rfloor.$$

This is easy if we think in binary.

With our new notations, Proposition 4.5 says that there exists an absolute constant $B > 0$ such that $\|e^{i\lambda(\delta)t}\|_{\delta/K} \geq B$, for every $K \subset I(\delta)$.

4.9 The second term

This term is really easy to bound. We have

$$\int_{I \smallsetminus I(x)} \frac{E_\alpha f(t)}{x - t} \, dt = \text{p.v.} \int_I \frac{E_\alpha f(t)}{x - t} \, dt - \text{p.v.} \int_{I(x)} \frac{E_\alpha f(t)}{x - t} \, dt.$$

Now these two intervals I and $I(x)$ are of the form that is used in definition 4.8 of the maximal Hilbert dyadic transform. Hence we have

$$\left| \int_{I \smallsetminus I(x)} \frac{E_\alpha f(t)}{x - t} \, dt \right| \leq 2 H_\alpha^* f(x). \tag{4.13}$$

4.10 The third term

We use here the fact that the numerator has a vanishing integral on every $J \in \Pi$. In this way we can change the first order singularity into a second order singularity, which is easier to handle.

Observe that by Proposition 4.7, $I \smallsetminus I(x)$ is the union of some members of the partition Π. Hence we can write

$$\int_{I \smallsetminus I(x)} \frac{e^{i\lambda(\alpha)(x-t)} f(t) - E_\alpha f(t)}{x - t} \, dt = \sum_{J_k} \int_{J_k} \frac{e^{i\lambda(\alpha)(x-t)} f(t) - E_\alpha f(t)}{x - t} \, dt,$$

where we are summing over those $J_k \in \Pi$ 'disjoint' from $I(x)$.

Let t_k be the center of J_k. We have

$$\frac{1}{x-t} - \frac{1}{x-t_k} = \frac{t-t_k}{(x-t)(x-t_k)}.$$

Now, the integral of $e^{i\lambda(\alpha)(x-t)}f(t) - E_\alpha f(t)$ on every J_k is zero, so that

$$\int_{I \smallsetminus I(x)} \frac{e^{i\lambda(\alpha)(x-t)}f(t) - E_\alpha f(t)}{x-t} dt =$$

$$\sum_{J_k} \int_{J_k} \frac{t-t_k}{(x-t)(x-t_k)} e^{i\lambda(\alpha)(x-t)} f(t)\, dt - \sum_{J_k} \int_{J_k} \frac{t-t_k}{(x-t)(x-t_k)} E_\alpha f(t)\, dt.$$

$$(4.14)$$

We want to reduce the third term to the function $\Delta_\alpha(x)$. First consider the second part of (4.14).

We recall that from Proposition 4.7 we know that $d(x, J_k) \geq \delta_k/2$, where $\delta_k = |J_k|$, for every J_k that appears in (4.14). Hence if $t \in J_k$, $\delta_k \leq |x - t_k|$ it follows that

$$|(x-t)(x-t_k)| \geq |x-t_k|^2 - \frac{\delta_k}{2}|x-t_k| \geq \frac{1}{2}|x-t_k|^2 \geq \frac{1}{4}|x-t_k|^2 + \frac{1}{4}\delta_k^2. \quad (4.15)$$

By definition $E_\alpha f$ is constant on every J_k. Hence we have

$$\left| \sum_{J_k} \int_{J_k} \frac{t-t_k}{(x-t)(x-t_k)} E_\alpha f(t)\, dt \right| \leq 2 \sum_{J_k} |E_\alpha f(t_k)| \int_{J_k} \frac{\delta_k\, dt}{(x-t_k)^2 + \delta_k^2}$$

$$= 2 \sum_{J_k} \frac{\delta_k^2}{(x-t_k)^2 + \delta_k^2} |E_\alpha f(t_k)|.$$

Now, as every term in the definition 4.9 of $\Delta(\Pi, x)$ is positive, we conclude that

$$\left| \sum_{J_k} \int_{J_k} \frac{t-t_k}{(x-t)(x-t_k)} E_\alpha f(t)\, dt \right| \leq 2 \sum_{J_k \in \Pi} \frac{\delta_k^2}{(x-t_k)^2 + \delta_k^2} |E_\alpha f(t_k)|. \quad (4.16)$$

We are now summing over all $J_k \in \Pi_\alpha$.

Now we bound the first term in (4.14). Here we use Theorem 4.3 again. The procedure is similar to that we have used to bound the first term. Let $\beta_k = \alpha/J_k$. We have

$$\sum_{J_k} \int_{J_k} \frac{t-t_k}{(x-t)(x-t_k)} e^{i\lambda(\alpha)(x-t)} f(t)\, dt$$

$$= \sum_{J_k} \frac{1}{|J_k|} \int_{J_k} \left\{ \frac{|J_k|(t-t_k)}{(x-t)(x-t_k)} e^{i(\lambda(\alpha) - \lambda(\beta_k))(x-t)} \right\} e^{i\lambda(\beta_k)(x-t)} f(t)\, dt.$$

Now we apply Theorem 4.3 to every integral. Applying (4.15)

$$\left\| \frac{|J_k|(t-t_k)}{(x-t)(x-t_k)} e^{i\left(\lambda(\alpha)-\lambda(\beta_k)\right)(x-t)} \right\|_\infty \le 2\frac{\delta_k^2}{(x-t_k)^2+\delta_k^2}.$$

The second derivative of

$$\frac{|J|(t-t_k)e^{iM(x-t)}}{(x-t)(x-t_k)}$$

is

$$-\frac{2|J|iMe^{iM(x-t)}}{(x-t)(x-t_k)} + \frac{2|J|e^{iM(x-t)}}{(x-t)^2(x-t_k)} - \frac{2|J|iM(t-t_k)e^{iM(x-t)}}{(x-t)^2(x-t_k)}$$
$$-\frac{|J|M^2(t-t_k)e^{iM(x-t)}}{(x-t)(x-t_k)} + \frac{2|J|(t-t_k)e^{iM(x-t)}}{(x-t)^3(x-t_k)}.$$

By (4.10), $\left||J|M\right| \le 2\pi$, and by (4.15) $|J_k| = \delta_k$, $|x-t| > \delta_k/2$ and $|t-t_k| < \delta_k/2$. We obtain that $\delta_k^2\|\varphi''\|_\infty$ is bounded by

$$[8\pi^2 + 32\pi + 32]\frac{\delta_k^2}{(x-t_k)^2+\delta_k^2}.$$

All this gives us

$$\left| \sum_{J_k} \int_{J_k} \frac{t-t_k}{(x-t)(x-t_k)} e^{i\lambda(\alpha)(x-t)} f(t)\, dt \right| \le C\sum_{J_k} \frac{\delta_k^2}{(x-t_k)^2+\delta_k^2}\, \|f\|_{\beta_k},$$
$$\tag{4.17}$$

where C is an absolute constant.

To compare (4.16) and (4.17) we observe that

$$|E_\alpha f(t_k)| \le C\|f\|_{\beta_k}. \tag{4.18}$$

In fact, we have that

$$|E_\alpha f(t_k)| = \frac{1}{|J_k|}\left| \int_{J_k} e^{i\lambda(\alpha)(x-t)} f(t)\, dt \right|$$
$$= \frac{1}{|J_k|}\left| \int_{J_k} e^{i\left(\lambda(\alpha)-\lambda(\beta_k)\right)(x-t)} e^{i\lambda(\beta_k)(x-t)} f(t)\, dt \right|.$$

We are now in position to apply Theorem 4.3. Here we use again (4.10) and finally obtain

$$|E_\alpha f(t_k)| \le C\|f\|_{\beta_k},$$

for some absolute constant C.

Therefore the third term in (4.5) is bounded by

$$D \sum_{J_k} \frac{\delta_k^2}{(x - t_k)^2 + \delta_k^2} \|f\|_{\beta_k}.$$

Now, as every summand in the definition of $\Delta_\alpha(x)$ is positive, we can write

$$|\text{third term in (4.5)}| \leq D \sup_{J_k} \|f\|_{\beta_k} \Delta_\alpha(x). \qquad (4.19)$$

4.11 First form of the basic step

Theorem 4.11 *Let $C_\alpha f(x)$ be a Carleson integral and $\Pi = \Pi_\alpha$ a dyadic partition of $I = I(\alpha)$. Assume that every $J \in \Pi$ has measure $\leq |I|/4$. Let $I(x)$ be the interval defined on proposition 4.7, and $\beta = \alpha/I(x)$. Then*

$$|C_\alpha f(x) - C_\beta f(x)| \leq C\|f\|_\beta + 2H_\alpha^* f(x) + D \sup_{J_k} \|f\|_{\alpha/J_k} \Delta_\alpha(x), \qquad (4.20)$$

where C and D are absolute constants.

Observe that by (4.9) we have

$$|C_\alpha f(x) - C_\beta f(x)| \leq C\|f\|_\beta + \sum_k \left| \int_{J_k} \frac{e^{i\lambda(\alpha)(x-t)} f(t)}{x - t} \, dt \right|.$$

We have introduced here the conditional expectation $E_\alpha f(x)$ and then we have replaced $(x-t)^{-1}$ by $(t-t_k)(x-t)^{-1}(x-t_k)^{-1}$. This is very convenient. For example, if we apply directly Theorem 4.3 we only obtain the bound $\leq C \sum_k \|f\|_{\alpha/J_k}$.

4.12 Some comments about the proof

1. In what sense can we say that $C_\beta f(x)$ is a 'simpler' Carleson Integral?

First, we can say that we pass from $C_\alpha f(x)$ to $C_\beta f(x)$ where if $n(\alpha) > 0$ then $n(\beta) \leq n(\alpha)$. Thus we can expect to obtain $n(\beta) = 0$. We can restrict the study to real functions, because, with f real, we have

$$C_{-\alpha} f(x) = \overline{C_\alpha f(x)}, \qquad (4.21)$$

if $\alpha = (n, I)$ and $-\alpha = (-n, I)$ with $n \in \mathbf{N}$. Hence we can assume that $n(\alpha)$ is positive.

Another case in which we can consider only values $n(\alpha) > 0$ is when we study functions f with $|f| = \chi_A$ (for some measurable set A). In this case $\overline{f}$ is of the same nature and

$$\mathcal{C}_{-\alpha}f(x) = \overline{\mathcal{C}_\alpha \overline{f}(x)}.$$

This is not all we have to say about question 1. But it is what we can say now.

2. We want to prove that the Carleson maximal operator is bounded from $\mathcal{L}^p(I)$ to $\mathcal{L}^p(I/2)$ $(1 < p < +\infty)$. But we can use the interpolation theorems to reduce the problem to proving the weak inequality

$$\mathfrak{m}\{\mathcal{C}_I^* f(x) > y\} \leq \frac{A_p \|f\|_p^p}{y^p}.$$

Hence we would like, given $f \in \mathcal{L}^p(I)$ and given $y > 0$, to define E with $\mathfrak{m}(E) < A_p \|f\|_p^p/y^p$ and such that for every $x \in I/2 \setminus E$ and $\alpha \in \mathcal{P}$ with $I(\alpha) = I$ we have $|\mathcal{C}_\alpha f(x)| < y$.

In fact, we will first construct, given $f \in \mathcal{L}^p(I)$, $y > 0$ and $N \in \mathbf{N}$, a subset E_N with $\mathfrak{m}(E_N) < A_p \|f\|_p^p/y^p$; such that for every $x \in I/2 \setminus E_N$ and $\alpha \in \mathcal{P}$ with $I(\alpha) = I$, and $0 \leq n(\alpha) < 2^N$ we will have $|\mathcal{C}_\alpha f(x)| < y$. Then

$$\{\mathcal{C}_I^* f > y\}$$
$$\subset \bigcup_N \{x \in I/2 : |\mathcal{C}_\alpha f(x)| > y, 0 \leq n(\alpha) < 2^N, I(\alpha) = I\},$$

and since $A_N = \{x \in I/2 : |\mathcal{C}_\alpha f(x)| > y, 0 \leq n(\alpha) < 2^N, I(\alpha) = I\}$ is an increasing sequence of sets, we will have

$$\mathfrak{m}(\{\mathcal{C}_I^* f > y\}) = \lim_N \mathfrak{m}(A_N) \leq A_p \|f\|_p^p/y^p.$$

Technical reasons will force us to replace 2^N by $\theta 2^N$ (θ an absolute constant) in the above reasoning.

3. Another important point about the proof is that we shall use the basic step repeatedly. Therefore, given a Carleson integral $\mathcal{C}_\alpha f(x)$ with $I(\alpha) = I$, $0 \leq n(\alpha) < 2^N$ and $x \notin E_N$, we shall obtain a sequence $(\alpha_j)_{j=1}^s$ in $\mathcal{P}$ with $\alpha_1 = \alpha$. Then we will have

$$|\mathcal{C}_\alpha f(x)| \leq \sum_{j=1}^{s-1} |\mathcal{C}_{\alpha_j} f(x) - \mathcal{C}_{\alpha_{j+1}} f(x)| + |\mathcal{C}_{\alpha_s} f(x)|. \tag{4.22}$$

We shall apply the basic step to every difference and will arrange things so that $n(\alpha_s) = 0$.

4.13 Choosing the partition Π_α. The norm $|||f|||_\alpha$

From now on we will consider a Carleson integral $\mathcal{C}_\alpha f(x)$ where $0 \leq n(\alpha) < 2^N$ and $I(\alpha) = J$, but we will not assume that $J = I$. If we want to apply the basic step to this integral, what selection of the partition Π will be good?

The intervals of the partition Π will be dyadic with respect to $J = I(\alpha)$ and of length less than $|J|/4$. What we want, in view of (4.20) is to have control of $\|f\|_{\alpha/J_k}$ for every $J_k \in \Pi$. Hence we put $b_j = 2 \cdot 2^{-2^j}$, and we will assume that for the intervals J_{00}, J_{01}, J_{10}, and J_{11} we have

$$\|f\|_{\alpha/J_{00}}, \|f\|_{\alpha/J_{01}}, \|f\|_{\alpha/J_{10}}, \|f\|_{\alpha/J_{11}} < y b_{j-1}. \tag{4.23}$$

We obtain the partition Π_α by a process of subdivision. We start with the four grandsons of $J = I(\alpha)$ of which we assume (4.23). Then at every stage of the process we take some interval K and we subdivide it in its two sons K_0 and K_1, if they satisfy the condition $\|f\|_{\alpha/K_0}, \|f\|_{\alpha/K_1} < y b_{j-1}$. If they do not, we consider K to be one of the intervals of the partition. Since this process can be infinite we also stop the division if $|K| \leq |I|/2^N$ and consider it to be of the partition.

As we need to consider the condition (4.23), we define for every $\alpha = (n, J) \in \mathcal{P}$

$$|||f|||_\alpha = \sup\{\|f\|_{\alpha/J_{00}}, \|f\|_{\alpha/J_{01}}, \|f\|_{\alpha/J_{10}}, \|f\|_{\alpha/J_{11}}\}.$$

It is important to see that this construction and the definition of $I(x)$ in proposition 4.7, imply that either $|I(x)| = 2|I|/2^N$ or $|||f|||_\beta \geq y b_{j-1}$.

Now we have another answer to the question 1. If we start with $y b_j \leq |||f|||_\alpha < y b_{j-1}$ we arrive at $|||f|||_\beta \geq y b_{j-1}$. We go from level j to a lesser level. It is true that a lesser level means here a greater norm $|||f|||_\beta$, but it also means a smaller number of cycles $n(\beta) \leq n(\alpha)$, and we will arrange things so that we arrive to $\mathcal{C}_{\alpha_s} f$ with $n(\alpha_s) = 0$. We also have a good bound of the differences $|\mathcal{C}_{\alpha_j} f(x) - \mathcal{C}_{\alpha_{j+1}} f(x)|$ in (4.22).

As we have motivated above, given $f \in \mathcal{L}^1(I)$ and $\alpha = (n, J) \in \mathcal{P}$, we define

$$|||f|||_\alpha = \sup\{\|f\|_{\alpha/J_{00}}, \|f\|_{\alpha/J_{01}}, \|f\|_{\alpha/J_{10}}, \|f\|_{\alpha/J_{11}}\}. \tag{4.24}$$

We will say a Carleson integral $\mathcal{C}_\alpha f(x)$ is of **level** $j \in \mathbf{N}$ if

$$y b_j \leq |||f|||_\alpha < y b_{j-1}.$$

In the construction that follows we assume $f \in \mathcal{L}^1(I)$, a Carleson integral $\mathcal{C}_\alpha f(x)$, a natural number N, and a real number $y > 0$ to be given; where $\alpha = (n, J)$ with $0 \leq n < 2^N$, and J being the union of two dyadic intervals with respect to I, has length $|J| > 4|I|/2^N$. We also assume that $\||f\||_\alpha < b_{j-1}y$ for some natural number j (not necessarily the level of $C_\alpha f(x)$).

Our objective is to select a convenient dyadic partition Π of J so that we can apply theorem 4.11.

We consider now the set of dyadic intervals J_u with respect to J, such that $|J|/4 \geq |J_u| \geq |I|/2^N$. For example, in the first four rows of the figure 4.3 we have represented these intervals.

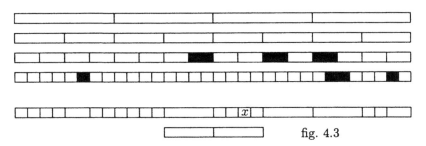

fig. 4.3

For every one of these intervals we determine if they satisfy the condition

$$\|f\|_{\alpha/J_u} < yb_{j-1}. \tag{4.25}$$

(In the figure we have painted in black the intervals that, hypothetically, do not satisfy this condition).

Now the interval J_u is a member of the partition Π if it is of length $|J_u| = |I|/2^N$ and it, all its ancestors and their brothers satisfy the condition (4.25), (as J_{10011} in the example of the figure) or it, all its ancestors and their brothers satisfy the condition (4.25) but one of its sons does not satisfy this condition (as J_{011} in the example). (The fifth row of the figure is a representation of the partition Π in the case we are handling).

Finally, observe that according to proposition 4.7, the interval $I(x)$ is always the union of one of the intervals of Π and a contiguous interval of the same length. Therefore $|I(x)| = 2|I|/2^N$ or some of the four grandsons of $I(x)$ will not satisfy the condition (4.25). (In the figure assuming that x is in the interval J_{10001}, the interval $I(x)$ is represented in the sixth row. In this case $I(x)$ is not a dyadic interval).

In our example, J_{10011} is a member of the partition Π, since its length is just $|I|/2^N$ and it, its ancestors J_{1001}, J_{100}, J_{10}, and their brothers J_{10010}, J_{1000}, J_{101}, J_{11} satisfy the condition (4.25). Also J_{011} is a member of the partition since it, its ancestor J_{01} and their brothers J_{010}, J_{00} satisfy the condition (4.25), but one of its sons does not, in this case J_{0111}.

4.14 Basic theorem, second form

Now that we have a good selection of the partition Π_α we can give a better version of the basic step. We shall need the following comparison between the two norms $\|\cdot\|_\alpha$ and $\|\|\cdot\|\|_\alpha$.

Proposition 4.12 *There is a constant $C > 0$ such that for every $f \in \mathcal{L}^2(J)$ and $\alpha = (n, J)$*

$$\|f\|_\alpha \leq C\|\|f\|\|_\alpha.$$

Proof. Let $|J| = \delta$, and denote by K the grandsons of J. Then

$$\|f\|_\alpha = \sum_j \frac{c}{1+j^2}\left|\frac{1}{\delta}\int_J f(t)\exp\left\{-2\pi i\left(n+\frac{j}{3}\right)\frac{t}{\delta}\right\}dt\right|.$$

Let $\delta' = \delta/4$. We can write

$$\|f\|_\alpha \leq \sum_K \sum_j \frac{c}{1+j^2}\frac{1}{4\delta'}\left|\int_K f(t)\exp\left\{-2\pi i\left(\frac{n}{4}+\frac{j}{12}\right)\frac{t}{\delta'}\right\}dt\right|,$$

where K denotes the grandsons of J. Let $n = 4m + r$ where $r = 0, 1, 2$ or 3 and put $4j + s$ instead of j, where $s = 0, 1, 2$ or 3. Then $\|f\|_\alpha$ is equal to

$$\sum_K \sum_s \sum_j \frac{c}{1+(4j+s)^2}\frac{1}{4\delta'}\left|\int_K f(t)\exp\left\{-2\pi i\left(m+\frac{j}{3}+\frac{r}{4}+\frac{s}{12}\right)\frac{t}{\delta'}\right\}dt\right|.$$

By Proposition 4.2 we have for $t \in K$

$$\exp\left\{2\pi i\left(\frac{r}{4}+\frac{s}{12}\right)\frac{t}{\delta'}\right\} = \sum_\ell c_\ell \exp\left(2\pi i\frac{\ell}{3}\frac{t}{\delta'}\right);$$

where $(1+\ell^2)|c_\ell| \leq B$ (and where c_ℓ depends on K, r, and s). Now we have $\|f\|_\alpha$ bounded by

$$\sum_K \sum_s \sum_j \frac{c}{1+(4j+s)^2}\sum_\ell \frac{|c_\ell|}{4\delta'}\left|\int_K f(t)\exp\left\{-2\pi i\left(m+\frac{j}{3}+\frac{\ell}{3}\right)\frac{t}{\delta'}\right\}dt\right|$$

$$\leq \sum_K \sum_s \sum_j \sum_\ell \frac{C}{1+j^2}|c_\ell||F(m,j+\ell,K)|$$

$$\leq C\sum_K \sum_{j,k} \frac{(1+k^2)}{(1+j^2)(1+(k-j)^2)}\frac{|F(m,k,K)|}{1+k^2}.$$

Now observe that

$$\frac{1+k^2}{(1+j^2)(1+(k-j)^2)} = \left(\frac{1}{1+j^2} + \frac{1}{1+(k-j)^2}\right)\frac{1+k^2}{2+j^2+(j-k)^2}$$

$$\leq 2\left(\frac{1}{1+j^2} + \frac{1}{1+(k-j)^2}\right).$$

Hence we have

$$\|f\|_\alpha \leq 2C \sum_K \sum_k \left(\sum_j \frac{1}{1+j^2} + \frac{1}{1+(k-j)^2}\right)\frac{|F(m,k,K)|}{1+k^2}$$

$$\leq D \sum_K \sum_k \frac{|F(m,k,K)|}{1+k^2}$$

$$\leq D \sum_K \|f\|_{\alpha/K} \leq 4D\|\|f\|\|_\alpha.$$

$\square$

In the same way we can prove that

$$\|f\|_\alpha \leq C \sup\{\|f\|_{\alpha/J_0}, \|f\|_{\alpha/J_1}\}, \qquad J = I(\alpha). \tag{4.26}$$

Now we can formulate the basic step with all its ingredients.

Theorem 4.13 (Basic Step) *Let $\xi \in \mathcal{P}_I$, and $x \in I(\xi)/2$ and assume that $\|\|f\|\|_\xi < yb_{j-1}$. Given $N \in \mathbf{N}$, let Π_ξ and $I(x)$ be the corresponding partition of $I(\xi)$ and interval, defined in Proposition 4.7. Let J be a smoothing interval such that $I(x) \subset J \subset I(\xi)$, and $x \in J/2$.*

Assume that $|\xi| \geq 4|I|/2^N$. Then we have

$$\left|\mathcal{C}_{\xi/J}f(x) - \mathcal{C}_{\xi/I(x)}f(x)\right| \leq Cyb_{j-1} + 2H_\xi^* f(x) + Dyb_{j-1}\Delta_\xi(x), \tag{4.27}$$

where C and D are absolute constants.

Proof. The condition $|\xi| \geq 4|I|/2^N$ assure us that we can apply the procedure to obtain Π_ξ.

We have seen that the selection of $I(x)$ implies that J is union of some members of Π_ξ, and by (4.26), $\|f\|_{\xi/J} \leq Cyb_{j-1}$. The same reasoning gives $\|f\|_{\xi/I(x)} \leq Cyb_{j-1}$.

Now observe that

$$\mathcal{C}_{\xi/J}f(x) = \text{p.v.} \int_J \frac{e^{2\pi i n(\xi/J)(x-t)/|J|}}{x-t} f(t)\, dt,$$

where $n(\xi/J) = \lfloor n(\xi)|J|/|I(\xi)|\rfloor$. By a change of frequency, we have

$$\left|\mathcal{C}_{\xi/J}f(x) - \text{p.v.} \int_J \frac{e^{2\pi i n(\xi)(x-t)/|I(\xi)|}}{x-t} f(t)\, dt\right| \leq B\|f\|_{\xi/J} \leq Cyb_{j-1}.$$

In spite of the possibility that $J \neq I(\xi)$ we can use the partition Π_ξ, as in the basic decomposition (4.1) for the integral

$$\text{p.v.} \int_J \frac{e^{2\pi i n(\xi)(x-t)/|I(\xi)|}}{x-t} f(t)\, dt. \tag{4.28}$$

We obtain a representation of the integral in (4.28) as

$$\text{p.v.} \int_{I(x)} \frac{e^{i\lambda(\xi)(x-t)}}{x-t} f(t)\, dt + \int_{J \smallsetminus I(x)} \frac{E_\xi f(t)}{x-t}\, dt$$
$$+ \int_{J \smallsetminus I(x)} \frac{e^{i\lambda(\xi)(x-t)} f(t) - E_\xi f(t)}{x-t}\, dt.$$

For the first term we obtain, as in (4.9), and by a change of frequency,

$$\mid \text{First term} - C_{\xi/I(x)} f(x) \mid \leq C \|f\|_{\xi/I(x)} \leq C y b_{j-1}.$$

The second can be bounded as in (4.13) by

$$\mid \text{Second term} \mid \leq 2 H_\xi^* f(x).$$

For the third term we must use the fact that $J \smallsetminus I(x)$ can be written as a union of intervals from Π_ξ. Then we proceed as in Theorem 4.11. In this case we obtain a sum of some of the terms of $\Delta_\xi(x)$ instead of all the terms. Since these terms are all positive this sum is less than $\Delta_\xi(x)$. Finally we obtain, as in (4.19),

$$\mid \text{Third term} \mid \leq D \sup_{J_k} \|f\|_{\beta_k} \Delta_\xi(x) \leq D y b_{j-1} \Delta_\xi(x).$$

This finishes the proof. $\qquad\qquad\qquad\qquad\qquad\qquad\qquad\qquad\qquad\qquad$ $\square$

5. Maximal Inequalities

In this chapter we give two inequalities to bound the two terms $\Delta_\xi(x)$ and $H_\xi^* f(x)$ that arise in the basic step.

5.1 Maximal inequality for $\Delta(\Pi, x)$

Theorem 5.1 *There are some absolute constants A and $B > 0$, such that for every finite partition Π of the interval $J \subset \mathbf{R}$ by intervals, we have*

$$\frac{\mathrm{m}\{x \in J : \Delta(\Pi, x) > y\}}{|J|} \le Ae^{-By}. \tag{5.1}$$

Proof. First recall that if the intervals J_k of Π have center at t_k and length δ_k we have defined the function $\Delta(\Pi, x)$ as

$$\Delta(\Pi, x) = \sum_k \frac{\delta_k^2}{(x - t_k)^2 + \delta_k^2}.$$

Let $g \colon \mathbf{R} \to [0, +\infty)$ be a bounded and measurable function. We can define a harmonic function on the upper-half plane by convolution with the Poisson kernel

$$u(x, y) = \frac{1}{\pi} \int_{\mathbf{R}} \frac{y\, g(t)}{(x - t)^2 + y^2}\, dt = P_y * g(x).$$

Hence we have

$$\int \Delta(\Pi, t)g(t)\, dt = \sum_k \pi \delta_k u(t_k, \delta_k) = \sum_k \pi \delta_k P_{\delta_k} * g(t_k).$$

If we assume g to be positive, Lemma 5.2 gives

$$P_{\delta_k} * g(t_k) \le \frac{2}{\delta_k} \int_{J_k} P_{\delta_k} * g(t)\, dt.$$

Therefore

$$\int \Delta(\Pi, t)g(t)\, dt \le 2\pi \sum_k \int_{J_k} P_{\delta_k} * g(t)\, dt.$$

By the general inequality of the Hardy-Littlewood maximal function (cf. Theorem 1.6), we get $P_{\delta_k} * g(t) \leq \mathcal{M}g(t)$. Hence

$$\int \Delta(\Pi, t) g(t) \, dt \leq 2\pi \int_J \mathcal{M}g(t) \, dt$$

Now, we have seen (cf. Proposition 1.5), that $\int_B \mathcal{M}f(x) \, dx \leq m(B) + 2c_1 \int |f(x)| \log^+ |f(x)| \, dx$; hence

$$\int \Delta(\Pi, t) g(t) \, dt \leq c|J| + c \int_{\mathbf{R}} |g(t)| \log^+ |g(t)| \, dt.$$

Now we put $g(t) = e^{y/2c} \chi_{\{\Delta(\Pi,t) > y\}}(t)$ and obtain

$$ye^{y/2c} m\{t : \Delta(\Pi, t) > y\} \leq \int \Delta(\Pi, t) g(t) \, dt$$

$$\leq c|J| + ce^{y/2c} \frac{y}{2c} m\{t : \Delta(\Pi, t) > y\}.$$

Hence

$$\frac{m\{t : \Delta(\Pi, t) > y\}}{|J|} \leq \frac{2c}{y} e^{-y/2c}.$$

Since the left member is less than or equal 1, we obtain the desired bound.

$\square$

We prove now the following lemma:

Lemma 5.2 *Let $g \in \mathcal{L}^\infty(\mathbf{R})$ be a positive function and $u(x, y) = P_y * g(x)$. Then for every interval $J \subset \mathbf{R}$ with center at a and length y we have*

$$u(a, y) \leq \frac{2}{|J|} \int_J u(x, y) \, dx.$$

Proof. Without loss of generality we can assume that $a = 0$. So we want to prove

$$\frac{1}{\pi} \int_{\mathbf{R}} \frac{y}{t^2 + y^2} g(-t) \, dt \leq \frac{2}{|J|} \int_J \frac{1}{\pi} \int_{\mathbf{R}} \frac{y}{t^2 + y^2} g(x - t) \, dt \, dx$$

$$= \frac{2}{|J|} \int_J \frac{1}{\pi} \int_{\mathbf{R}} \frac{y}{(x + t)^2 + y^2} g(-t) \, dt \, dx.$$

But this follows from

$$\frac{y}{t^2 + y^2} \leq \frac{2}{|J|} \int_J \frac{y}{(x + t)^2 + y^2} \, dx = \frac{2}{y} \int_{-y/2}^{y/2} \frac{y}{(x + t)^2 + y^2} \, dx. \qquad (5.2)$$

Changing the variable we see that (5.2) is equivalent to

$$\frac{1}{u^2+1} \leq 2 \int_{-1/2}^{1/2} \frac{dx}{(x+u)^2+1}.$$

And we see that for every ξ with $|\xi| < 1/2$ we have

$$\frac{1}{u^2+1} \leq \frac{2}{(\xi+u)^2+1}.$$

□

5.2 Maximal inequality for $H_I^* f$

In order to prove an inequality of the same type as (5.1) for $H_I^* f$ we first relate this maximal function to the ordinary maximal Hilbert transform $\mathcal{H}^* f$ and the maximal function of Hardy and Littlewood.

Proposition 5.3 *Let $1 \leq p < +\infty$, $f \in \mathcal{L}^p(\mathbf{R})$, and let $I \subset \mathbf{R}$ be a bounded interval. Then for every $x \in I$ we have*

$$H_I^* f(x) \leq 2\mathcal{H}^* f(x) + 6\mathcal{M} f(x). \tag{5.3}$$

Proof. Observe that in the definition of $H_I^* f(x)$ we consider only the restriction of f to the interval I.

Let $K \subset I$ be such that $x \in K/2$. Let $J \subset K$ be an interval with center at x and of maximal length, and $L \supset K$ be an interval with center at x but of minimal length. We can write

$$\left| \int_K \frac{f(t)}{x-t} dt \right| \leq \left| \int_J \frac{f(t)}{x-t} dt \right| + \left| \int_{K \smallsetminus J} \frac{f(t)}{x-t} dt \right|.$$

The first term is equal to $\mathcal{H} f(x) - \int_{\mathbf{R} \smallsetminus J} \cdots$ and consequently is bounded by $2\mathcal{H}^* f(x)$.

To bound the second term observe that if $t \in K \smallsetminus J$, then $|x - t| \geq d(x, \mathbf{R} \smallsetminus K) \geq |K|/4$. Hence

$$\left| \int_{K \smallsetminus J} \frac{f(t)}{x-t} dt \right| \leq \frac{4}{|K|} \int_{K \smallsetminus J} |f(t)| \, dt \leq \frac{4|L|}{|K|} \frac{1}{|L|} \int_L |f(t)| \, dt.$$

From $x \in K/2$ and the definition of L follows that $|L| \leq 6|K|/4$. Hence we have obtained (5.3). □

Theorem 5.4 *There are absolute constants A and $B > 0$ such that for every $f \in \mathcal{L}^\infty(I)$, and every $y > 0$*

$$\frac{\mathbf{m}\{H_I^* f(x) > y\}}{|I|} \leq A e^{-By/\|f\|_\infty}. \tag{5.4}$$

Proof. By homogeneity we can assume that $\|f\|_\infty = 1$. We shall also consider f as the restriction to I of a function vanishing on $\mathbf{R} \setminus I$.

By proposition 5.3 as $\|\mathcal{M}f\|_\infty \le 1$ we have

$$\{H_I^* f(x) > y\} \subset \{2\mathcal{H}^* f(x) > y/2\},$$

if $y > 12$.

Now we want to bound

$$\int_I e^{A\mathcal{H}^* f(t)} \, dt.$$

We have seen that the maximal Hilbert transform satisfies

$$\|\mathcal{H}^* f\|_p \le Bp\|f\|_p, \qquad 2 < p < +\infty.$$

Hence

$$\int_I \left(e^{A\mathcal{H}^* f(t)} - A\mathcal{H}^* f(t)\right) dt \le |I| + \sum_2^{+\infty} \frac{A^n}{n!} B^n n^n |I|.$$

If we choose A small enough (A only depending on B), we arrive at

$$\int_I \left(e^{A\mathcal{H}^* f(t)} - A\mathcal{H}^* f(t)\right) dt \le C|I|.$$

It follows that

$$\mathfrak{m}\{\mathcal{H}^* f(x) > y/4\}\left(e^{Ay/4} - Ay/4\right) \le C|I|.$$

Therefore, if $y > y_0$ (y_0 depending only on A), we have

$$\frac{1}{2} e^{Ay/4} \mathfrak{m}\{\mathcal{H}_I^* f(x) > y\} \le C|I|.$$

Hence we have proved (5.4) for $y > y_0$. Now, changing A, we can forget the restriction on y. $\qquad\square$

6. Growth of Partial Sums

6.1 Introduction

As a first indication that the basic step is powerful we give a theorem about the partial sums of the Fourier series of a function $f \in \mathcal{L}^2[0, 2\pi]$. This can be regarded as a toy example of the techniques involved in Carleson's theorem. This example will also justify our procedure in the proof of Carleson's Theorem.

The principal result in this chapter will be that if $f \in \mathcal{L}^2[0, 2\pi]$, then $S_n(f, x) = o(\log \log n)$ almost everywhere. As we have seen, to bound $S_n(f, x)$ we must bound the corresponding Carleson integral. Hence our first objective will be to bound $\sup_\alpha |\mathcal{C}_\alpha f(x)|$ where the supremum is taken over all pairs α with $I(\alpha) = I$ and $|n(\alpha)| < \theta 2^N$. Each time we apply the basic step we lessen the values of $n(\alpha)$ and $|\alpha|$; the role of θ is to assure that we arrive to $n(\alpha) = 0$ before we arrive to $|\alpha| < 4|I|/2^N$. In this chapter $\theta = 1/4$, but later, in the proof of the Carleson Theorem, we shall need another value of θ.

We will consider that f is real, so that by (4.21) we can assume $0 \leq n(\alpha) < \theta 2^N$.

Given $\varepsilon > 0$, we shall construct a set $E = \bigcup_{N=2}^\infty E_N \subset I$ with $\mathfrak{m}(E) < A\varepsilon$ and such that

$$\sup_{\substack{0 \leq n(\alpha) < \theta 2^N \\ I(\alpha) = I}} |\mathcal{C}_\alpha f(x)| < B \frac{\|f\|_2}{\sqrt{\varepsilon}} \log N, \qquad \text{for } x \in I/2 \smallsetminus E.$$

To achieve this we will use the procedure indicated in (4.22). Every piece E_N of the exceptional set will be the union of four subsets $E_N = S \cup T_N \cup U_N \cup V$.

T_N and U_N will allow us to bound the terms $H_\xi^* f(x)$ and $\Delta_\xi(x)$ when we apply the basic step.

V will allow us to bound the final term $|\mathcal{C}_{\alpha_s} f(x)|$ in (4.22) when $n(\alpha_s) = 0$.

The definition of S will be given so that for every $x \in I/2 \smallsetminus S$ the integrals $\mathcal{C}_{\alpha_j} f(x)$ that appear in (4.22) will have a level $j \geq 1$.

Let $\mathcal{C}$ be the set of pairs (n, J), where $0 \leq n < \theta 2^N$, and J is a smoothing interval with respect to I, such that $|J| \geq 4|I|/2^N$.

6.2 The seven trick

We are going to define the exceptional set E_N. Given $\alpha \in \mathcal{C}$, from $I(\alpha) \not\subset E_N$, we want to derive some properties of every grandchild J of $I(\alpha)$. Therefore, for every such property p, we shall define the set

$$A = \bigcup \{J : J \text{ dyadic and } p(J)\},$$

where $p(J)$ denotes that J satisfies the property we are considering.

Now, given a measurable set A, we can define the set

$$A^* = \bigcup \{7J : J \subset A, \text{ dyadic}\},$$

where $7J$ denotes the interval of length $7|J|$ and the same center as J.

Hence for every dyadic interval $J \subset A$ we put in A^* the interval J and three contiguous intervals of the same length at each side.

Then the subset A^* satisfies our conditions. In fact if $I(\alpha) \not\subset A^*$ and J is a grandchild of $I(\alpha)$ we have $J \not\subset A$. Suppose that $J \subset A$. Since J is dyadic we would have $I(\alpha) \subset 7J \subset A^*$, that is, a contradiction.

Moreover, two dyadic intervals are disjoint or one is contained in the other, and all dyadic intervals are subsets of I. Hence we can obtain a set $\mathcal{V}$ of disjoint dyadic intervals $J \subset A$ such that

$$A^* = \bigcup_{J \in \mathcal{V}} 7J.$$

Therefore

$$\mathrm{m}(A^*) \leq \sum_{J \in \mathcal{V}} 7|J| \leq 7\mathrm{m}(A).$$

And in general we have that if J is a smoothing interval such that $J \not\subset A^*$, then every grandchild K of J satisfies $K \not\subset A$.

6.3 The exceptional set

Let $y > 0$. Later we will determine y big enough so that $\mathrm{m}(E) < A\varepsilon$.

The first component of the exceptional set is $S^* = \bigcup_J 7J$, where the union is taken for all the dyadic intervals, J, such that

$$\frac{1}{|J|} \int_J |f|^2 \, dm \geq y^2. \tag{6.1}$$

The definition of S^* is given so that:
For every $\alpha \in \mathcal{P}_I$ such that $x \in I(\alpha)/2$, but $x \notin S^*$ we have $\||f\||_\alpha \leq y$.
In fact, for every grandson J of $I(\alpha)$, we have

$$\frac{1}{|J|} \int_J |f|^2 \, dm < y^2.$$

(In the other case $x \in I(\alpha) \subset 7J \subset S^*$). Therefore for every grandson J of $I(\alpha)$ we have $\|f\|_{\alpha/J} < y$. Hence $\||f\||_\alpha < y$.

Two dyadic intervals are disjoint or one is contained in the other. It follows that there exists a sequence (J_n) of disjoint dyadic intervals, such that every J_n satisfies (6.1), and every other J that satisfies (6.1) is contained in one of the J_n. Hence

$$m(S^*) = m\left(\bigcup_n 7J_n\right) \leq 7 \sum_n m(J_n) \leq \frac{7}{y^2} \int_I |f|^2 \, dm \leq \frac{7}{y^2} \|f\|_2^2.$$

Let $\alpha \in \mathcal{C}$ such that $I(\alpha) \not\subset S^*$, then $\||f\||_\alpha < y$. Hence $\||f\||_\alpha = 0$ or there exists $j \in \mathbf{N}$ such that $yb_j \leq \||f\||_\alpha < yb_{j-1}$. With α, j, y and N we obtain a partition Π_α of $I(\alpha)$ as in section [4.13]. Now, for such α, we define $T_N(\alpha) = \emptyset$ if $\||f\||_\alpha = 0$ or, in other case,

$$T_N(\alpha) = \{x \in I(\alpha) : H_\alpha^* f(x) > Myb_{j-1} \log(\sqrt{N}/b_j)\}, \qquad (6.2)$$

where $M > 0$ and $y > 0$ will be determined later. Observe also that $\sqrt{N} > b_j$ always.

The norm $\|E_\alpha f\|_\infty$ has been bounded in (4.18) in terms of the local norms of f on the intervals of the partition, hence we have the bound $\|E_\alpha f\|_\infty \leq C \sup_k \|f\|_{\beta_k} < Cyb_{j-1}$. Therefore by the maximal inequality obtained in (5.4) for the maximal Hilbert dyadic transform,

$$m(T_N(\alpha)) \leq A|\alpha| \exp\left(-\frac{BM}{C} \log(\sqrt{N}/b_j)\right)$$

$$\leq A|\alpha| \frac{b_j^2}{N^3} \leq A|\alpha| \frac{\||f\||_\alpha^2}{y^2 N^3},$$

if we choose M in such a way that $BM/C = 6$.

Let T_N be the union of the sets $T_N(\alpha)$. To bound $m(T_N)$ we need the following:

Lemma 6.1 *For every $f \in \mathcal{L}^2(J)$*

$$\sum_{n \in \mathbf{Z}} \||f\||_{\alpha_n}^2 \leq 16\left(\frac{1}{|J|} \int_J |f|^2 \, dm\right),$$

where $\alpha_n = (n, J)$ for every $n \in \mathbf{Z}$.

Now we have

$$m(T_N) = \sum_{\alpha=(n,J)} m(T_N(\alpha)) \leq \sum_J \sum_n A|J|y^{-2}N^{-3}\|\|f\|\|_\alpha^2$$

$$\leq 16Ay^{-2}N^{-3}\sum_J \int_J |f|^2\, dm.$$

Here J can be any smoothing interval of I whose length $\geq 4|I|/2^N$. Summing first for all J with the same length $|I|/2^r$,

$$m(T_N) \leq 16Ay^{-2}N^{-3}\sum_{r=0}^{N-2} 2\int_I |f|^2\, dm \leq \frac{32A}{y^2N^2}\|f\|_2^2. \qquad (6.3)$$

The set U_N will also be the union of $U_N(\alpha)$, for the same α's. Put

$$U_N(\alpha) = \{x \in I(\alpha) : \Delta_\alpha(x) > (M/C)\log(\sqrt{N}/b_j)\},$$

where M, C and y are the same constants that appear in (6.2). By the corresponding maximal inequality of theorem 5.1 we obtain:

$$m(U_N) = \sum_J \sum_n m(U_N(\alpha)) \leq \sum_J \sum_n A|J|y^{-2}N^{-3}\|\|f\|\|_\alpha^2.$$

The same reasoning we used before proves the inequality

$$m(U_N) \leq \frac{32A}{y^2N^2}\|f\|_2^2.$$

Finally, the last component V will be the set

$$V = \{x \in I : H_I^*f(x) > y\}$$

By the relation, proved in proposition 5.3, between the maximal Hilbert dyadic transform, the maximal Hilbert transform and the maximal Hardy-Littlewood function, the function H^*f is of weak type $(2,2)$. Therefore we have

$$m(V) \leq C\frac{\|f\|_2^2}{y^2}.$$

Observe that given $f \in \mathcal{L}^2(I)$, $N \in \mathbf{N}$, and $y > 0$, we have constructed a measurable set $E_N = S^* \cup T_N \cup U_N \cup V \subset I$ such that for $E = \bigcup_{N=2}^\infty E_N$ we have

$$m(E) = m(S^*) + m(V) + \sum_{N=2}^\infty m(T_N \cup U_N) \leq A\frac{\|f\|_2^2}{y^2},$$

where A denotes an absolute constant.

Proof of lemma 6.1. First, we prove the inequality

$$\sum_{n\in\mathbf{Z}} \|f\|_{\alpha_n}^2 \le \Big(\frac{1}{|J|}\int_J |f|^2\,dm\Big).$$

There exists $(x_n)_{n\in\mathbf{Z}} \in \ell^2(\mathbf{Z})$ with $\sum_n |x_n|^2 = 1$ such that

$$\Big(\sum_{n\in\mathbf{Z}} \|f\|_{\alpha_n}^2\Big)^{1/2} = \sum_{j,n} \frac{x_n}{1+j^2}\Big|\frac{c}{|J|}\int_J f(t)\exp\Big\{-2\pi i\Big(n+\frac{j}{3}\Big)\frac{t}{|J|}\Big\}\,dt\Big|.$$

Hence if we denote by y_{3n+j} the above integral, we have

$$\sum_{n\in\mathbf{Z}} |y_{3n}|^2 = \sum_{n\in\mathbf{Z}} |y_{3n+1}|^2 = \sum_{n\in\mathbf{Z}} |y_{3n+2}|^2 = |J|^{-1}\int_J |f(t)|^2\,dt.$$

Therefore

$$\Big(\sum_{n\in\mathbf{Z}} \|f\|_{\alpha_n}^2\Big)^{1/2} = c\sum_{j,n}\frac{x_n y_{3n+j}}{1+j^2} \le \|f\|_{\mathcal{L}^2(J)}\sum_j \frac{c}{1+j^2} = \|f\|_{\mathcal{L}^2(J)}.$$

Now we denote by K the grandsons of J and we have

$$\sum_{n\in\mathbf{Z}} \||f\||_{\alpha_n}^2 \le \sum_K \sum_n \|f\|_{(\lfloor n/4\rfloor, K)}^2 \le 4\sum_K \sum_m \|f\|_{(m,K)}^2$$

$$\le 4\sum_K \frac{1}{|K|}\int_K |f|^2\,dm \le \frac{16}{|J|}\sum_K \int_K |f|^2\,dm \le \frac{16}{|J|}\int_J |f|^2\,dm.$$

$\square$

6.4 Bound for the partial sums

Proposition 6.2 *Let $f \in \mathcal{L}^2(I)$, and $\varepsilon > 0$ be given. There exists a set E of measure $\mathrm{m}(E) < A\varepsilon$, such that for every $N > 2$ and $x \in I/2 \smallsetminus E$*

$$\sup_{0\le k<\theta 2^N} |\mathcal{C}_{(k,I)}f(x)| \le B\frac{\|f\|_2}{\sqrt{\varepsilon}}(\log N). \tag{6.4}$$

Proof. Without loss of generality we can assume that f is a real function with $\|f\|_2 > 0$. Choosing $y = \|f\|_2/\sqrt{\varepsilon}$, we have constructed in the previous section a set E such that $\mathrm{m}(E) < A\varepsilon$.

Consider the set $\mathcal{C}'$ of those pairs α such that $I(\alpha)$ is a smoothing interval of length $|\alpha| > 4|I|/2^N$ and $0 \le n(\alpha)|I|/|\alpha| < \theta 2^N$. (This is a subset of the set of pairs $\mathcal{C}$ that we have used in section 1). The set $\mathcal{C}'$ is defined in order that for every pair $\alpha = (k, I)$ with $0 \le k < \theta 2^N$ and every $J \subset I$ a smoothing interval of length $> 4|I|/2^N$ we have $\alpha/J \in \mathcal{C}'$.

Now choose $x \in I/2 \setminus E$. For every Carleson integral $\mathcal{C}_\alpha f(x)$ appearing in (6.4), we have $\alpha \in \mathcal{C}'$.

Assume that we have a Carleson integral $\mathcal{C}_\alpha f(x)$ with $\alpha \in \mathcal{C}'$. Since $x \notin S^*$ and $x \in I(\alpha)/2$, we have $\|\|f\|\|_\alpha < y$. Hence there is a well defined level $j \in \mathbf{N}$ such that $b_j y \le \|\|f\|\|_\alpha < y b_{j-1}$. (Or $f = 0$ a. e. on $I(\alpha)$ and there is nothing to prove).

We are in position to apply the procedure of section [4.13] to obtain a dyadic partition Π_α of $I(\alpha)$ with intervals J of length $|I|/2^N \le |J| \le |I|/4$. Then Proposition 4.7 gives us a smoothing interval $I(x)$. This interval $I(x)$ is the union of an interval J_0 of Π_α and a contiguous interval. If $|J_0| = |I|/2^N$, then $|I(x)| = 2|I|/2^N$. Otherwise there is a son K of J_0 (hence a grandson of $I(x)$) such that $\|f\|_{\alpha/K} \ge y b_{j-1}$. Therefore we have $|I(x)| = 2|I|/2^N$ or $\|\|f\|\|_{\alpha/I(x)} \ge y b_{j-1}$.

Put $\beta = \alpha/I(x)$. We are going to prove that we have a good bound for $|\mathcal{C}_\alpha f(x) - \mathcal{C}_\beta f(x)|$. Also either $\beta \in \mathcal{C}'$ or we have a good bound for $\mathcal{C}_\beta f(x)$.

When $|I(x)| \le 4|I|/2^N$ we have $n(\beta) = 0$. In fact,

$$n(\beta) = \left\lfloor n(\alpha) \frac{|I(x)|}{|\alpha|} \right\rfloor = \left\lfloor n(\alpha) \frac{|I|}{|\alpha|} \frac{|I(x)|}{|I|} \right\rfloor < \theta 2^N \frac{4}{2^N} = 1.$$

Since $x \notin V$ we will have $|\mathcal{C}_\beta f(x)| \le y$, a bound that is good enough for us.

In the second case $|I(x)| = |I(\beta)| > 4|I|/2^N$ and

$$n(\beta) \frac{|I|}{|\beta|} = \left\lfloor n(\alpha) \frac{|I(x)|}{|\alpha|} \right\rfloor \frac{|I|}{|\beta|} \le n(\alpha) \frac{|I|}{|\alpha|} < \theta 2^N.$$

Therefore $\beta \in \mathcal{C}'$. Also by Proposition 4.7, $x \in I(\beta)/2$. If $\|\|f\|\|_\beta = 0$, we obtain $\mathcal{C}_\beta f(x) = 0$ and so we have a good bound of $\mathcal{C}_\beta f(x)$. Otherwise there exists $k \in \mathbf{N}$ such that $y b_k \le \|\|f\|\|_\beta < y b_{k-1}$. The construction of Π_α and the fact that $\|\|f\|\|_\beta \ge y b_{j-1}$ imply that $j > k$.

The basic step gives us in all cases (taking $J = I(\alpha)$)

$$|\mathcal{C}_\alpha f(x) - \mathcal{C}_\beta f(x)| \le C y b_{j-1} + 2 H_\alpha^* f(x) + D y b_{j-1} \Delta_\alpha(x).$$

Now since $x \notin E_N$ this is

$$\le C y b_{j-1} + 2 M y b_{j-1} \log(\sqrt{N}/b_j) + D y b_{j-1}(M/C) \log(\sqrt{N}/b_j)$$

$$\le C y b_{j-1} \log(\sqrt{N}/b_j).$$

Now either $\mathcal{C}_\beta f(x) = 0$, or $n(\beta) = 0$, or we are in position to apply the same procedure with β instead of α. In the last case we also have a level $1 \le k < j$, so that in a finite number of steps we must arrive at $n(\beta) = 0$ or $\mathcal{C}_\beta f(x) = 0$.

Then we obtain

$$|\mathcal{C}_\alpha f(x)| \le y + \sum_{j \in \mathbf{N}} C y b_{j-1} \log(\sqrt{N}/b_j) \le B(\log N) y.$$

$\square$

Proposition 6.3 *Let $f \in \mathcal{L}^2(I)$, then*

$$\sup_{0 \leq |k| < n} |\mathcal{C}_{(k,I)}f(x)| = o(\log \log n), \qquad \text{a.e. on } I/2.$$

Proof. As we did before, we assume that f is a real function. The proposition is equivalent to

$$\sup_{0 \leq k < \theta 2^N} |\mathcal{C}_{(k,I)}f(x)| = o(\log N), \qquad \text{a.e. on } I/2. \qquad (6.5)$$

What we have proved can be written as

$$\limsup_{N \to +\infty} \frac{\sup_{0 \leq k < \theta 2^N} |\mathcal{C}_{(k,I)}f(x)|}{\log N} \leq B \frac{\|f\|_2}{\sqrt{\varepsilon}}, \qquad x \notin E.$$

Now we need the fact (cf. Proposition 4.1) that for every trigonometric polynomial P, there exists a constant $C(P) < +\infty$ such that

$$\sup_{0 \leq k < \theta 2^N} |\mathcal{C}_{(k,I)}P(x)| \leq C(P).$$

Hence there exists a set E', with $\mathrm{m}(E') < A\varepsilon$, such that for $x \notin E'$

$$\limsup_{N \to +\infty} \frac{\sup_{0 \leq k < \theta 2^N} |\mathcal{C}_{(k,I)}f(x)|}{\log N}$$

$$= \limsup_{N \to +\infty} \frac{\sup_{0 \leq k < \theta 2^N} |\mathcal{C}_{(k,I)}(f - P)(x)|}{\log N} \leq B \frac{\|f - P\|_2}{\sqrt{\varepsilon}}.$$

The density of the trigonometric polynomial can be used now to prove that the set of points where

$$\limsup_{N \to +\infty} \frac{\sup_{0 \leq k < \theta 2^N} |\mathcal{C}_{(k,I)}f(x)|}{\log N} > 0$$

is of measure $\leq A\varepsilon$. $\qquad \square$

Proposition 6.4 *Let $f \in \mathcal{L}^2[-\pi, \pi]$, and let $S_n(f, x)$ be the partial sums of the Fourier series of f. Then*

$$S_n(f, x) = o(\log \log n), \qquad \text{a.e. on } [-\pi, \pi].$$

Proof. The partial sums are given by the convolution of f with the Dirichlet kernel $S_n(f, x) = D_n * f(x)$. Thus, by the expression (2.3) of the Dirichlet kernel we get

$$S_n(f, x) = \frac{1}{\pi} \int_{-\pi}^{\pi} f^\circ(x - t) \frac{\sin t}{t} \, dt + \frac{1}{2\pi} \int_{-\pi}^{\pi} f^\circ(x - t) \varphi_n(t) \, dt,$$

for every x with $|x| < \pi$, and where $\|\varphi_n\|_\infty \le C$ uniformly on $n \in \mathbf{N}$, and f° denotes the periodic extension of f for $|t| < 2\pi$ and 0 for $|t| \ge 2\pi$. Therefore as $\sin t/t \in \mathcal{L}^2(\mathbf{R})$ we get

$$\left| S_n(f,x) - \frac{1}{\pi} \int_{-2\pi}^{2\pi} f^\circ(t) \frac{\sin(x-t)}{x-t}\, dt \right| \le C\|f\|_2.$$

Thus

$$\sup_{0 \le n \le N} |S_n(f,x)| \le \sup_{0 \le |n| \le N} |\mathcal{C}_\alpha f^\circ(x)| + C\|f\|_2,$$

where the supremum on the right is taken over those α with $I(\alpha) = [-2\pi, 2\pi]$.

By proposition 6.3 it follows that

$$\sup_{0 \le n \le N} |S_n(f,x)| = o(\log\log N).$$

$\square$

Remark. Obviously Proposition 6.4 is superseded by Carleson's Theorem. Carleson proved instead that $S_n(f,x) = o(\log\log n)$ a.e. if there exists $\delta > 0$ such that $\int_I |f|(\log^+ |f|)^{1+\delta}\, d\mathfrak{m} < +\infty$. The proof is almost the same but Lemma 6.1 must be replaced by an inequality of Hausdorff-Young type

$$\sum_{n \in \mathbf{Z}} \exp\left(-a\|f\|_{\alpha_n}^{-1/(1+\delta)}\right) < +\infty,$$

valid whenever $\int_I |f|(\log^+ |f|)^{1+\delta}\, d\mathfrak{m} < +\infty$.

At the time of the publication of the Carleson theorem the best result known about $S_n(f,x)$ for $f \in \mathcal{L}^2(I)$ was the Kolmogorov-Seliverstov-Plessner theorem giving $S_n(f,x) = o(\sqrt{\log n})$ a.e.

7. Carleson analysis of the function

7.1 Introduction

Carleson's Theorem is achieved by refining the proof in the previous chapter. The weak point of this proof is that we have allowed all the pairs $\alpha \in \mathcal{C}'$, and for each one a fraction of $I(\alpha)$ must be included in E_N, $(T_N(\alpha) \cup U_N(\alpha)$ where we have not a good bound for the second and third term in the basic decomposition). Since we are summing over all pairs $\alpha = (n, J)$ with J any smoothing interval of I with length $\geq 4|I|/2^N$ we obtain a factor N in $m(T_N)$ and $m(U_N)$ (see equation (6.3)). We are forced to compensate it by a term $\log(\sqrt{N})$ in the exponent. Finally this logarithmic term appears in the final bound of the Carleson integral.

But it is clear from the proof that not all pairs are used in the inductive steps. So we must define a set of **allowed pairs** and assure that we only use these at the inductive steps.

There is also the possibility that we change the frequency of a Carleson integral, introducing a controllable error, to transfer it to an allowed pair.

It is clear that possible candidates for the allowed pairs will be those intervals that appear as $I(x)$ in the process of choosing the partition in Proposition 6.2. In this process we check for every dyadic interval below $I(\alpha)$ if the condition $\|f\|_{\alpha/J_u} < yb_{j-1}$ is satisfied. The intervals $I(x)$ are selected between the grandfathers of those that do not satisfy this condition.

Recalling that $\|f\|_\beta$ are mean values of generalized Fourier coefficients of f on the interval $I(\beta)$, we are persuaded to think that the allowed pairs must be related to large local Fourier coefficients.

This, maybe, justifies the following step in Carleson's construction: what we will call the **Carleson analysis of the function** f. This can be seen as the process of writing the score from a piece of music. The following section, that can be skipped, tries to explain this connection.

In this and the following chapter we assume a function $f \in \mathcal{L}^2(I) \cap \mathcal{L}^p(I)$ to be given, where $1 \leq p < +\infty$. Later we shall assume that $|f| = \chi_A$ is the characteristic function of a measurable set of I, and we shall use interpolation techniques to recover the case of a general $f \in \mathcal{L}^p(I)$.

7.2 A musical interlude

When we are hearing music our ears and mind are working very heavily. What do we hear? A possible answer to this question is that our mind analyzes the sound signal f to obtain the notes that compose f. These notes are what we sense as music. Obviously this is a good answer: when a musician wants to preserve the music he writes the score, and what he writes must be a substantial portion of what we hear.

If we look at the text of a musical composition we see connected dots situated on five parallel lines. The vertical axis represents the pitch, the horizontal axis the time. Every dot represents a note. We can think of a note as a wave train $ae_\alpha(x)$. The human sense of pitch is determined by the frequency of vibration of the sound. The frequency of $ae_\alpha(x)$ is $\lambda(\alpha)$. Other elements of the note are its intensity, represented by the constant a, its duration $|\alpha|$, and the interval of time in which the note must be produced, $I(\alpha)$. Hence every dot is associated with a pair $\alpha = (n, I)$, and the music can be seen as a set of pairs α that are the notes that compose the sound. If the music has a binary rhythm, such as one determined by a time signature of 2 over 2 or 4 over 4, the notes will be on dyadic intervals, such as the one we are dealing with in the proof of Carleson's Theorem. The signal f can be written in this case as

$$f(x) = \sum_{j=1}^{h} a_j e_{\alpha_j}(x).$$

In real music a note has a quality of tone that is not captured in the pair α. In fact the ear senses a note as opposed to noise if the sound signal f is periodic. During a fraction of a second (say a fraction greater that $1/64$ of a second) the graph of f is periodic with a periodicity between 27 (the note A_{-2}) and 4176 (the note C_7) cycles by second. But the function is not a sinusoidal function. It has a complicated Fourier coefficients structure. And we are considering only its period.

In the text of a musical composition the articulations, accents, and nuances of the tone strength are designated at the best in a very faulty way and often not at all, so that the musical interpretation has an important role. This implies more or less that it is not easy to deduce our coefficients a_j, and other harmonic notes from the score. This makes our previous expression for f very poor.

For our purpose what is important is: The allowed pairs that we must select to obtain a proof of Carleson's Theorem are connected with the **notes** that compose the function f.

We must obtain a device that given the sound f gives us the notes that compose the music f. This is what we will call the Carleson analysis of the function f.

7.3 The notes of f

The allowed pairs must be determined so that they contain pairs for which $\||f\||_\alpha$ is specially great. To this purpose we carry out the *Carleson analysis of the function f*.

We fix a level j and retain the pairs $\alpha = (n, I)$ for which the corresponding Fourier coefficient of f is greater that $b_j y^{p/2}$. Next we develop the rest of the function on the two sons of I. We retain also those terms with coefficients greater that $b_j y^{p/2}$, and so on. We will call these pairs the **notes of f** at level j. The set of allowed pairs will be a modification of these.

In what follows we need to consider only pairs (n, J) where J is a dyadic interval. We define $\mathcal{D}_I$ as this set:

$$\mathcal{D}_I = \{\alpha \in \mathcal{P} : I(\alpha) \text{ a dyadic interval with respect to } I\}.$$

If $\alpha = (n, J)$ and J is a dyadic interval with respect to I, then there is a unique $u \in \{0, 1\}^*$ such that $J = I_u$. We put $u = u(\alpha)$.

In general if $u \in \{0, 1\}^*$ we call u' his father. For instance $0100110' = 010011$. Hence if $\alpha = (n, J)$ and J is a dyadic interval, $I_{u'(\alpha)}$ is the father of J.

The structure of f will determine sets of pairs $\mathcal{Q}_u^j$ that we call notes of f at level j. These pairs contain the information about f that we shall need.

We are going to define, by induction, for every level $j \in \mathbf{N}$ and every $u \in \{0, 1\}^*$, a set $\mathcal{Q}_u^j$ and a function $P_u^j(x)$. In the musical interpretation this function represents the sound of the notes of level j that last for all of the interval I_u.

For technical reasons we must take $\mathcal{Q}_\emptyset^j = \mathcal{Q}_0^j = \mathcal{Q}_1^j = \emptyset$ and the corresponding functions $P_\emptyset^j = P_0^j = P_1^j = 0$

In the first step of the induction we define for every u of length 2, that is $u = 00$, or $u = 01$, or $u = 10$, or $u = 11$:

$$\mathcal{Q}_u^j = \{\alpha \in \mathcal{D}_I : I(\alpha) = I_u, |\langle f, e_\alpha \rangle| \geq b_j y^{p/2} |I_u|\}. \tag{7.1}$$

That is to say, we retain the pairs α that give a Fourier coefficient greater than or equal to $b_j y^{p/2}$, where $b_j = 2 \cdot 2^{-2^j}$, and define

$$P_u^j(x) = \sum_{\alpha \in \mathcal{Q}_u^j} \frac{\langle f, e_\alpha \rangle}{|I_u|} e_\alpha(x); \tag{7.2}$$

and, assuming we have defined $\mathcal{Q}_u^j$ and P_u^j, and that $v = u0$ or $v = u1$ we define

$$\mathcal{Q}_v^j = \{\alpha \in \mathcal{D}_I : I(\alpha) = I_v, |\langle f - P_u^j, e_\alpha \rangle| \geq b_j y^{p/2} |I_v|\}, \tag{7.3}$$

$$P_v^j(x) = P_u^j(x) + \sum_{\alpha \in \mathcal{Q}_v^j} \frac{\langle f - P_u^j, e_\alpha \rangle}{|I_v|} e_\alpha(x). \tag{7.4}$$

We define the set of notes of level j as the union $\mathcal{Q}^j = \bigcup_{u \in \{0,1\}^*} \mathcal{Q}_u^j$.

With the above definition it is easy to see that

$$P_u^j(x) = \sum_{\substack{\alpha \in \mathcal{Q}^j \\ I(\alpha) \supset I_u}} \frac{\langle f - P_{u'(\alpha)}^j, e_\alpha \rangle}{|\alpha|} e_\alpha(x) = \sum_{\substack{\alpha \in \mathcal{Q}^j \\ I(\alpha) \supset I_u}} a(\alpha) e_\alpha(x). \tag{7.5}$$

Observe that these coefficients $a(\alpha)$ also depend on j.

By definition $P_v^j - P_u^j$ is orthogonal to $f - P_v^j$ on I_v, when $v = u0$ (or $v = u1$). Therefore

$$\int_{I_v} |f(x) - P_v^j(x)|^2 \, dx + \sum_{\alpha \in \mathcal{Q}_v^j} |a(\alpha)|^2 |\alpha| = \int_{I_v} |f(x) - P_u^j(x)|^2 \, dx.$$

Summing over all the dyadic intervals of length $|v| \leq n$,

$$\sum_{|u|=n} \int_{I_u} |f(x) - P_u^j(x)|^2 \, dx + \sum_{\substack{\alpha \in \mathcal{Q}^j \\ |u(\alpha)| \leq n}} |a(\alpha)|^2 |\alpha| = \int_I |f(x)|^2 \, dx.$$

Therefore

$$\sum_{\alpha \in \mathcal{Q}^j} |a(\alpha)|^2 |\alpha| \leq \int_I |f(x)|^2 \, dx. \tag{7.6}$$

The definition of $a(\alpha)$ implies that for every $\alpha \in \mathcal{Q}^j$, $|a(\alpha)| \geq b_j y^{p/2}$. Hence the length of all the pairs in $\mathcal{Q}^j$ is bounded by

$$\sum_{\alpha \in \mathcal{Q}^j} |\alpha| \leq b_j^{-2} \frac{\|f\|_2^2}{y^p}. \tag{7.7}$$

This is the **bound of the length of the notes of level** j.

7.4 The set X

Here we define a component of the exceptional set. The objective is that when $I_u \not\subset X$ we have a good bound for $P_u^j(x)$, and also a bound for the number of terms in the sum that defines this function.

Equation (7.6) invites us to define the function

$$A_j(x) = \sum_{\alpha \in \mathcal{Q}^j} |a(\alpha)|^2 \, \chi_{I(\alpha)}(x).$$

In some way this function represents the intensity of the sounds x.

By (7.6), we have

$$\int_I A_j(x)\, dx \le \|f\|_2^2.$$

For every j we define the set

$$X_j = \{x \in I : A_j(x) > y^p/b_j\}.$$

By definition, A_j is a union of a set of dyadic intervals. Chebyshev's inequality gives us the bound

$$\mathfrak{m}(X_j) \le b_j \frac{\|f\|_2^2}{y^p}.$$

Now we put

$$X = \bigcup_j X_j.$$

Therefore

$$\mathfrak{m}(X) \le C \frac{\|f\|_2^2}{y^p}.$$

Proposition 7.1 *Let I_u be a dyadic interval, such that $I_u \not\subset X$. Then P_u^j has at most b_j^{-3} terms and*

$$|P_u^j(x)| \le \sum_{I(\alpha) \supset I_u, \alpha \in \mathcal{Q}^j} |a(\alpha)| \le \frac{y^{p/2}}{b_j^2}. \tag{7.8}$$

Proof. Notice that the condition $I_u \not\subset X$ implies that $I_u \not\subset X_j$. Hence there exists a point $x_0 \in I_u$ such that $x_0 \notin X_j$. So

$$A_j(x_0) = \sum_{\alpha \in \mathcal{Q}^j} |a(\alpha)|^2 \, \chi_{I(\alpha)}(x_0) \le y^p/b_j.$$

For every term $a(\alpha)e_\alpha(x)$ of P_u^j, by (7.5), $x_0 \in I(\alpha)$. Denote by N the number of terms in P_u^j, and notice that $|a(\alpha)| \ge b_j y^{p/2}$. From the previous inequality we deduce

$$Nb_j^2 y^p \le y^p/b_j.$$

This is the bound on the number of terms of P_u^j.

Now

$$|P_u^j(x)| \le \sum_{\substack{\alpha \in Q^j \\ I(\alpha) \supset I_u}} |a(\alpha)||e_\alpha(x)| \le \sqrt{N}\left(\sum |a(\alpha)|^2\right)^{1/2}$$

$$\le b_j^{-3/2} A_j(x_0)^{1/2} \le b_j^{-3/2} y^{p/2} b_j^{-1/2} = b_j^{-2} y^{p/2}.$$

The set X is a union of dyadic intervals. We define X^* following the seven trick so that

$$m(X^*) \le C\frac{\|f\|_2^2}{y^p}. \tag{7.9}$$

Therefore if $J \not\subset X^*$ then $K \not\subset X$ for every grandson K of J. $\qquad\square$

7.5 The set S

Besides X^*, the set S^* will be a component of the exceptional set E_N. The definition is almost the same as in the case of the chapter on partial growth of the sums of Fourier series, but here we have another element, the number p.

> The definition of S is given so that every pair α with $I(\alpha) \not\subset S^*$ satisfies $\||f|\|_\alpha < y$.

Hence, at every point in which we are interested, $C_\alpha f(x)$ will have a well defined level or $\||f|\|_\alpha = 0$ in which case $C_\alpha f(x) = 0$ and we have achieved our objective: to bound $C_\alpha f(x)$.

We define S as the union of all the intervals J, where J is dyadic with respect to I, and

$$\frac{1}{|J|} \int_J |f|^p \, dm \ge y^p. \tag{7.10}$$

Hence for every dyadic interval that satisfies (7.10) we put in S the interval J.

Now if $\alpha = (n, J) \in \mathcal{P}_I$ and $J \not\subset S^*$, we will have for every grandson J_u of J

$$\frac{1}{|J_u|} \int_{J_u} |f| \, dm \le \left(\frac{1}{|J_u|} \int_{J_u} |f|^p \, dm\right)^{1/p} < y.$$

Hence $\|f\|_{\alpha/J_u} < y$, and

$$\||f|\|_\alpha < y. \tag{7.11}$$

Now if $x \notin S^*$, $\alpha = (n, J) \in \mathcal{P}_I$ and $x \in I(\alpha)/2$, we will also have $I(\alpha) \not\subset S^*$, and $\|\|f\|\|_\alpha < y$.

As a consequence of (7.11), if $I(\alpha) \not\subset S^*$, $\alpha \in \mathcal{P}_I$, the Carleson integral $\mathcal{C}_\alpha f(x)$ has level $j \in \mathbf{N}$, that is,

$$yb_j \leq \|\|f\|\|_\alpha < yb_{j-1}, \tag{7.12}$$

or $\|\|f\|\|_\alpha = 0$, which implies $f = 0$ on $I(\alpha)$.

The set S can be written as the union $\bigcup_k J_k$ where every J_k is a maximal dyadic interval satisfying (7.10). The intervals J_k are 'disjoint', hence we have

$$\mathfrak{m}(S) \leq \sum_k \mathfrak{m}(J_k),$$

where every J_k satisfies (7.10). Therefore

$$\mathfrak{m}(S) \leq \sum |J_k| \leq \frac{1}{y^p} \sum_k \int_{J_k} |f|^p \, dm \leq \frac{1}{y^p} \|f\|_p^p.$$

This is part of the bound of the exceptional set

$$\mathfrak{m}(S^*) \leq \frac{7}{y^p} \|f\|_p^p. \tag{7.13}$$

8. Allowed Pairs

8.1 The length of the notes

We need a set of pairs $\mathcal{S}$, the allowed pairs, so that in order to bound a given Carleson integral $\mathcal{C}_\alpha f(x)$ we apply repeatedly the basic decomposition, but always to integrals $\mathcal{C}_\beta f(x)$ with β being an allowed pair. In each application we may change the integral in question $\mathcal{C}_\alpha f(x)$ to another $\mathcal{C}_\beta f(x)$ if the difference $|\mathcal{C}_\alpha f(x) - \mathcal{C}_\beta f(x)|$ can be conveniently bounded. The advantage of using the allowed pairs is that for every pair in which we use the basic decomposition we must exclude a certain set, where we can not bound the second and third terms of the decomposition. These exceptional sets are a proportion of the sets $I(\beta)$. Hence the set $\mathcal{S}$ is subject to the restriction

$$\sum_{\beta \in \mathcal{S}} |\beta| < +\infty. \tag{8.1}$$

We have seen that a possible candidate for $\mathcal{S}$ is the set of notes (on every level) of f. This crude selection must be refined, and this is the subject of this chapter. The general idea is that we can extend the set of notes, always restricted by (8.1), in order to achieve that every Carleson integral $\mathcal{C}_\alpha f(x)$ can be approximated by a Carleson integral $\mathcal{C}_\beta f(x)$ with $\beta \in \mathcal{S}$.

Hence, now a basic condition is to verify that the set of notes has its total length bounded. Assuming that $f \in \mathcal{L}^2(I)$ we have obtained this bound in (7.7).

The condition $f \in \mathcal{L}^2(I)$ is used here to apply Parseval's Inequality. This is the principal obstacle to the $\mathcal{L}^p$ result. Carleson had the feeling that this can be removed but finally it was Hunt who obtained this extension. We will assume now that $f \in \mathcal{L}^2(I) \cap \mathcal{L}^p(I)$; at a certain point we will also assume that $|f|$ is a characteristic function. Later an interpolation argument will give the general result.

8.2 Well situated notes

We are going to define the set of well situated pairs. These pairs are an enlargement of the set $\mathcal{Q} = \bigcup_{j \geq 1} \mathcal{Q}^j$ of notes of f. We add pairs that in a certain sense are near the notes of f. This will define the set $\mathcal{R}$ of well situated notes or pairs. This set is defined as the sets of pairs that satisfy one of two conditions.

Our objective is that given a Carleson integral $\mathcal{C}_\alpha f(x)$ there exists an **allowed** β near α. If $P_{u(\alpha)}^j$ consists of a single note γ we have a good candidate $\beta = \gamma/\alpha$ for this β. A good candidate must also be a note with a comparable pitch. So we enlarge the set of notes of f adding all the notes for which there are two or more simultaneous notes of f of comparable pitch, because if there are two or more simultaneous notes we will not have a unique candidate for β. This will be no problem if the lengths of all the added notes are controlled.

Therefore the aim of the first stage of the definition is that from $\alpha \notin \mathcal{R}^j$ it follows that the function $P_{u(\alpha)}^j(x)$, on the interval $I(\alpha)$, essentially consists of a single note or a rest. When we speak of essentially we are refering to the notes with a comparable pitch to the note α. How can we achieve this? If there are two notes δ and γ in $\mathcal{Q}^j$ comparable to α, we add to $\mathcal{R}^j$ all the comparable notes on the intervals contained in $I(\delta) \cap I(\gamma)$. In this way we achieve that such an α, for which there exist δ and γ, will be in $\mathcal{R}^j$.

The pair $\alpha \in \mathcal{D}_I$ will be an element of $\mathcal{R}^j$ if it satisfies one of two conditions $A(\mathcal{R}^j)$ or $B(\mathcal{R}^j)$. For the benefit of those who want to read the original paper of Carleson we retain in these notations the reference to the letters a and b. First we give only condition $B(\mathcal{R}^j)$ and prove that it achieves our purpose: to obtain a precise information about the function $P_{u(\alpha)}^j$ when $\alpha \notin \mathcal{R}^j$.

Condition $B(\mathcal{R}^j)$ Let $\alpha \in \mathcal{D}_I$. If there is $\beta \in \mathcal{Q}^j$ such that $I(\alpha) \subset I(\beta)$ then, by definition $\alpha \in \mathcal{R}^j$ if there are two different elements γ and δ in $\mathcal{Q}^j$ with $I(\gamma) \cap I(\delta) \supset I(\beta)$ and such that

$$b_j^{10} \leq |\lambda(\gamma) - \lambda(\delta)| \cdot |\alpha| \leq 32 \cdot b_j^{-10} \quad \text{and} \quad |n(\alpha) - n(\gamma/\alpha)| < b_j^{-10}.$$

The role of $I(\beta)$ in this condition is to simplify the calculation of $\sum |\alpha|$ for those α that satisfy the condition.

As we have said this condition is designed so that the following lemma can be proved.

Lemma 8.1 (Structure of the functions P_u^j) Let $\alpha \in \mathcal{D}_I$ be such that $I(\alpha) \not\subset X$. Assume that $\alpha \notin \mathcal{R}^j$ and put $u = u(\alpha)$. Then we can write $P = P_u^j$ as

$$P(t) = \rho e^{i\lambda(\delta)t} + P_0(t) + P_1(t), \qquad t \in I(\alpha) \tag{8.2}$$

where $P_1(t)$ consists of the terms of $P(t)$ for which $|n(\alpha) - n(\gamma/\alpha)| \geq b_j^{-10}$, and

$$|\rho| \leq b_j^{-2} y^{p/2}, \quad \text{and} \quad |P_0(t)| \leq b_j^8 y^{p/2}, \qquad \text{if } t \in I(\alpha). \qquad (8.3)$$

Furthermore, if $\rho \neq 0$, then $\delta \in \mathcal{Q}^j$ and satisfies $I(\delta) \supset I(\alpha)$.

Proof. By (7.5),

$$P(t) = \sum_{I(\gamma) \supset I(\alpha), \gamma \in \mathcal{Q}^j} a(\gamma) e_\gamma(t) = {\sum_\gamma}' \cdots + {\sum_\gamma}'' \cdots$$

where the ${\sum_\gamma}'$ refers to those terms with $|n(\alpha) - n(\gamma/\alpha)| < b_j^{-10}$. Therefore by definition

$$P_1(t) = {\sum_\gamma}'' \cdots$$

satisfies the conditions of the theorem.

Now if the sum ${\sum_\gamma}'$ is empty, we put $\rho = 0$ and $P_0(t) = 0$ and the lemma is true. If the sum is reduced to one term $a(\delta) e_\delta(t)$ we can put $\rho = a(\delta)$ and $P_0(t) = 0$. Since $I_u \not\subset X$, (7.8) gives us $|\rho| \leq b_j^{-2} y^{p/2}$.

Finally if there are two or more elements in the sum ${\sum_\gamma}'$ we must have for every two of them γ and δ that

$$|\lambda(\gamma) - \lambda(\delta)| \cdot |\alpha| < b_j^{10}. \qquad (8.4)$$

In the other case we can assume without loss of generality that $I(\gamma) \subset I(\delta)$ (they are dyadic intervals), and take $\beta = \gamma$. Then α satisfies the condition $B(\mathcal{R}^j)$. In fact,

$$|\lambda(\gamma) - \lambda(\delta)| \cdot |\alpha| = 2\pi \left| n(\gamma) \frac{|\alpha|}{|\gamma|} - n(\delta) \frac{|\alpha|}{|\delta|} \right|.$$

By definition $n(\gamma/\alpha) = \lfloor n(\gamma)|\alpha|/|\gamma| \rfloor$ and by the construction of γ and δ we have $|n(\alpha) - n(\gamma/\alpha)| < b_j^{-10}$ and $|n(\alpha) - n(\delta/\alpha)| < b_j^{-10}$. Hence

$$|\lambda(\gamma) - \lambda(\delta)| \cdot |\alpha| \leq 2\pi(|n(\gamma/\alpha) - n(\delta/\alpha)| + 2) \leq 2\pi(2b_j^{-10} + 2) \leq 32 \cdot b_j^{-10}.$$

As we know by hypothesis that $\alpha \notin \mathcal{R}^j$ we have proved (8.4).

We choose δ to be one of the terms in the sum ${\sum_\gamma}'$. Let t_0 be the central point of $I(\alpha)$, we have for every $t \in I(\alpha)$ (Observe that for every γ that appears in the sum below we have $t \in I(\gamma)$)

$$\sum_\gamma{}' a(\gamma) e_\gamma(t) = \left(\sum_\gamma{}' a(\gamma) e^{i\lambda(\gamma) t_0} \right) e^{i\lambda(\delta)(t - t_0)} +$$

$$\sum_\gamma{}' a(\gamma) e^{i\lambda(\gamma) t_0} \left[e^{i\lambda(\gamma)(t - t_0)} - e^{i\lambda(\delta)(t - t_0)} \right]$$

$$= \rho e^{i\lambda(\delta) t} + P_0(t).$$

Again (7.8) gives us that $|\rho| \le b_j^{-2} y^{p/2}$.

On the other hand

$$|P_0(t)| \le \sum_\gamma' |a(\gamma)| \left| e^{i\left(\lambda(\gamma) - \lambda(\delta)\right)(t-t_0)} - 1 \right|.$$

Since $t \in I(\alpha)$ and t_0 is the central point of this interval, we have

$$\left| (\lambda(\gamma) - \lambda(\delta))(t-t_0) \right| \le \left| \lambda(\gamma) - \lambda(\delta) \right| \cdot |\alpha| \le b_j^{10}.$$

Hence, since $I(\alpha) \not\subset X$

$$|P_0(t)| \le b_j^{10} b_j^{-2} y^{p/2} = b_j^8 y^{p/2}.$$

$\square$

Remark. In the case that there exists $\delta \in \mathcal{Q}^j$ with $I(\delta) \supset I(\alpha)$ and $|n(\alpha) - n(\delta/\alpha)| < b_j^{-10}$, we can choose $\lambda(\delta)$ as the exponent that appears in (8.2).

The next condition in the definition of $\mathcal{R}^j$ is added in order to achieve, under certain hypotheses, that the function $P_u^j(x)$ coincides on every grandson I_v of I_u with $P_v^j(x)$.

Condition $A(\mathcal{R}^j)$ *If $\beta \in \mathcal{Q}^j$ then $\alpha \in \mathcal{R}^j$ for every $\alpha \in \mathcal{D}_I$ such that*

$$\begin{aligned} I(\alpha) \subset I(\beta), \quad &|\alpha| > b_j^{10} |\beta|, \\ \text{and} \quad &|n(\alpha) - n(\gamma/\alpha)| < b_j^{-10}. \end{aligned} \tag{8.5}$$

where $\gamma \in \mathcal{Q}^j$ is such that $I(\gamma) \supset I(\beta)$.

This condition needs also an exceptional set that allows the reasoning of Lemma (8.2).

For every $\alpha \in \mathcal{Q}^j$ let $Y_j(\alpha)$ be the union of the two intervals of length $8b_j^3 |\alpha|$ with centers at the two extremes of $I(\alpha)$. Then we set

$$Y = \bigcup_{j=1}^\infty \bigcup_{\alpha \in \mathcal{Q}^j} Y_j(\alpha). \tag{8.6}$$

Hence we have, by the bound (7.7) of the length of the notes of level j,

$$\mathfrak{m}(Y) \le \sum_{j=1}^\infty \sum_{\alpha \in \mathcal{Q}^j} 16 b_j^3 |\alpha| \le 16 \sum_{j=1}^\infty b_j \frac{\|f\|_2^2}{y^p} \le \frac{C}{y^p} \|f\|_2^2. \tag{8.7}$$

Lemma 8.2 *Let $\alpha \in \mathcal{P}_I$ such that $\alpha/L \notin \mathcal{R}^j$ for every grandchild L of $I(\alpha)$, and $I(\alpha) \not\subset Y$. Let P_u^j and P_v^j be the functions associated to two grandchildren $J = I_u$ and $K = I_v$ of $I(\alpha)$. If there is a term $a(\gamma)e_\gamma(x)$ of P_u^j, such that*

$$|n(\gamma/J) - n(\alpha/J)| < b_j^{-9},$$

then $P_u^j = P_v^j$. Hence the four functions associated to the grandchildren of α are the same.

Proof. The hypotheses imply that $\gamma \in \mathcal{Q}^j$ is such that $I(\gamma) \supset J$. Since $\alpha/J \notin \mathcal{R}^j$ by condition $A(\mathcal{R}^j)$ (taking $\beta = \gamma$) we conclude that $|\alpha/J| \leq b_j^{10}|\gamma|$. It follows that the size of $I(\gamma)$ is very large compared with that of $I(\alpha)$.

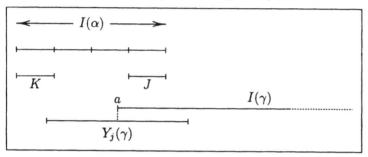

Fig. 1

Now we shall show that $I(\gamma) \supset I(\alpha)$. To this purpose we must show that it contains every grandchildren of $I(\alpha)$, say K. In fact if $I(\gamma) \not\supset K$ there is one end point a of $I(\gamma)$ contained in $I(\alpha)$. $Y_j(\gamma)$ contains an interval with center at a and length

$$8b_j^3|\gamma| > 8b_j^{10}|\gamma| \geq 2|\alpha|.$$

Hence it must be that $I(\alpha) \subset Y_j(\gamma) \subset Y$. This contradicts our hypothesis. Therefore $I(\gamma)$ contains every grandchild of $I(\alpha)$ and hence $I(\gamma) \supset I(\alpha)$.

From the above reasoning we can not only deduce that $I(\gamma) \supset I(\alpha)$ but also that the distance ϱ from $I(\alpha)$ to the end points of $I(\gamma)$ must satisfy

$$\varrho + |\alpha| \geq 4b_j^3|\gamma|.$$

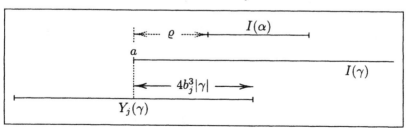

Fig. 2

It follows that
$$\varrho \geq 4(b_j^3 - b_j^{10})|\gamma| > b_j^7|\gamma|.$$

In order to prove that the two functions
$$P_u^j(x) = \sum_{\delta \in \mathcal{Q}^j, I(\delta) \supset J} a(\delta)e_\delta(x), \quad \text{and} \quad P_v^j(x) = \sum_{\delta \in \mathcal{Q}^j, I(\delta) \supset K} a(\delta)e_\delta(x),$$

coincide, we must prove that every $\delta \in \mathcal{Q}^j$ such that $I(\delta) \supset J$ satisfies also $I(\delta) \supset K$. (Observe that now we have proved $I(\gamma) \supset I(\alpha)$, and there is no special assumption about the interval J).

Hence we assume that $\beta \in \mathcal{Q}^j$, $I(\beta) \supset J$ and $I(\beta) \not\supset K$. It follows that one end point b of $I(\beta)$ is contained in $I(\alpha)$. Since $Y_j(\beta) \not\supset I(\alpha)$ it follows that $8b_j^3|\beta| < 2|\alpha|$. Now we can prove that $I(\gamma) \supset I(\beta)$. In fact, $I(\gamma)$ contains an interval of length $|\alpha| + 2\varrho$ with the same center as $I(\alpha)$. Hence all we have to show is that $\varrho > |\beta|$. But we have

$$\varrho > b_j^7|\gamma| > \frac{1}{4}b_j^{-3}|\alpha| > |\beta|.$$

We can apply now condition $A(\mathcal{R}^j)$, with our β and γ, to prove that $\alpha/J \in \mathcal{R}^j$. This contradicts our hypotheses. Hence it follows that $I(\beta) \supset K$, and the two functions P_u^j and P_v^j coincide. $\square$

Remark. There is nothing magical about the exponent 9. This will be applied in Proposition 9.3 and any number between 1 and 10 suffices.

8.3 The length of well situated notes

We must bound the length of the well situated notes. In fact we bound only the well situated pairs α such that $I(\alpha) \not\subset X$. The importance of this condition is that assuming it we can apply Proposition 7.1. That is, we know that $P_{u(\alpha)}^j$ has at most b_j^{-3} terms.

First observe that given $\beta \in \mathcal{Q}^j$, the number of cycles $n(\alpha)$ of a pair α that satisfies condition $A(\mathcal{R}^j)$ with this β is less than $2b_j^{-10}$ times the number k of $\gamma \in \mathcal{Q}^j$ such that $I(\gamma) \supset I(\beta)$. Since $I(\alpha) \subset I(\beta)$ we have $I(\beta) \not\subset X$. Therefore $k \leq b_j^{-3}$. Independently, the intervals $I(\alpha)$ must be dyadic subintervals of $I(\beta)$ of length between $b_j^{10}|\beta|$ and $|\beta|$. The length of all these intervals is

$$\sum_{I(\alpha) \subset I(\beta), |\alpha| \geq b_j^{10}|\beta|} |\alpha| \leq (1 + \log_2 b_j^{-10})|\beta|.$$

Therefore, by the known estimates of the length of the notes of f (7.7), with j fixed,

$$\sum_{\alpha \in A(\mathcal{R}^j), I(\alpha) \not\subset X} |\alpha| \leq \sum_{\beta \in \mathcal{Q}^j} C b_j^{-13} (\log b_j^{-1}) |\beta|$$

$$\leq C b_j^{-15} (\log b_j^{-1}) \frac{\|f\|_2^2}{y^p} \leq C b_j^{-16} \frac{\|f\|_2^2}{y^p}.$$

In the same way, the number of pairs γ and δ, that with a fixed β, satisfies conditions $B(\mathcal{R}^j)$ is $\leq b_j^{-3} \cdot b_j^{-3}$. (As in the previous case we are interested only in the case $I(\alpha) \not\subset X$). With these γ and δ fixed, the number of cycles $n(\alpha)$ can be selected between $2 b_j^{-10}$ values, and the length of $|\alpha|$ can attain only $\leq C(\log b_j^{-1})$ values. With these observations we can conclude that

$$\sum_{\alpha \in B(\mathcal{R}^j), I(\alpha) \not\subset X} |\alpha| \leq C b_j^{-6} b_j^{-10} (\log b_j^{-1}) b_j^{-2} \frac{\|f\|_2^2}{y^p} \leq C b_j^{-19} \frac{\|f\|_2^2}{y^p}.$$

Therefore the **lengths of the well situated notes** of level j with $I(\alpha) \not\subset X$ is bounded in the following way:

$$\sum_{\alpha \in \mathcal{R}^j, I(\alpha) \not\subset X} |\alpha| \leq C b_j^{-19} \frac{\|f\|_2^2}{y^p}. \tag{8.8}$$

8.4 Allowed pairs

Recall the basic step of the proof. We have a Carleson integral $\mathcal{C}_\alpha f(x)$, $I(\alpha) \not\subset S^*$ and this gives us that the Carleson integral has a well defined level $j \in \mathbf{N}$ such that

$$y b_j \leq \|\|f\|\|_\alpha < y b_{j-1}.$$

We apply the process by which we obtain a partition Π_α, and to obtain a good bound for the terms of the decomposition we carefully select a smoothing interval $I(x)$. Now, in general, one of the halves of $I(x)$ is an interval of Π_α. This implies that a grandson J of $I(x)$ satisfies

$$\|f\|_{\alpha/J} \geq y b_{j-1}. \tag{8.9}$$

Now we pass from $\mathcal{C}_\alpha f(x)$ to $\mathcal{C}_\beta f(x)$ where $\beta = \alpha/I(x)$.

We can assume that α was an allowed pair, but the problem is that in general β is not such a pair. The idea of the proof is to choose the set of allowed pairs $\mathcal{S}^j$ so that we can choose $\gamma \in \mathcal{S}^j$ with $|\mathcal{C}_\beta f(x) - \mathcal{C}_\gamma f(x)|$ conveniently bounded.

We have obtained a set $\mathcal{R}^j$ such that $\delta \notin \mathcal{R}^j$ implies that $P_{u(\delta)}^j$ sounds as a rest or a pure note ξ if we consider only the sounds of comparable pitch to that of δ. By (8.9) for some grandchildren J of $I(\beta)$, $\|f\|_{\beta/J} \geq y b_{j-1}$. This implies that certain coefficients of Fourier of f on J with a pitch comparable to that of β are of a certain size. The hope is that taking $\delta = \beta/J$ we can arrange things so that we can in fact prove that there exists (in $P^j_{u(\delta)}$) the note ξ of this pitch. Then the candidate to γ will be this note $\xi/I(x)$.

According to the previous considerations we must define $\beta \notin S^j$ so that it implies $\beta/J \notin R^j$. This will not be sufficient because we can not prove the implication

$$\|f\|_{\beta/J} \geq yb_{j-1} \implies \left\{ \begin{array}{l} \text{there exists one note in } P^j_{u(\beta/J)} \\ \text{of pitch near to that of } \beta. \end{array} \right.$$

Instead we shall prove that there exists a natural number m only depending on p, and such that

$$\|f\|_{\beta/J} \geq yb_{j-1} \implies \left\{ \begin{array}{l} \text{there exists one note in } P^{m+j}_{u(\beta/J)} \\ \text{of pitch near to that of } \beta. \end{array} \right.$$

This shift is of no consequence to the rest of the proof. In fact the natural number m must be chosen great enough to attain also another objective. For the present we take m depending only on p and great enough.

Then we can define S^j, the set of allowed pairs:

Definition 8.3 (Allowed pairs) A pair α is in S^j if $I(\alpha) \not\subset X^*$ is a smoothing interval, and for some grandchild J of $I(\alpha)$ we have $\alpha/J \in R^{m+j}$.

8.5 The exceptional set

Here we define the exceptional set. We assume $1 < p < +\infty$, the function $f \in \mathcal{L}^p(I) \cap \mathcal{L}^2(I)$, a positive real number $y > 0$ and a natural number N to be given. And our purpose is to define a set $E_N \subset \mathbf{R}$, with $m(E_N) < A\|f\|_p/y^p$ such that for every $x \in I/2 \smallsetminus E_N$ and $\alpha = (n, I) \in \mathcal{P}$ with $0 \leq n < 2^N$, we have $|\mathcal{C}_\alpha f(x)| < B_p y$.

In fact we will assume that $|f| = \chi_A$ is the characteristic function of a measurable set. Therefore once we obtain a bound of $|\mathcal{C}_{(n,I)}f(x)|$ for $0 \leq n < 2^N$, we shall have also a bound for $0 \leq |n| < 2^N$ applying the same reasoning to $\overline{f}$.

The set E_N, and other auxiliary sets that we will define later, will depend heavily on f, p, y and N. We will not mention the dependence on f, p, and y, but we shall mention the dependence on p of every constant that appears. Recall that the shift m depends on p and only on p.

The set E_N will be a union $D \cup S^* \cup T_N \cup U_N \cup V \cup X^* \cup Y$ of various sets.

We retain N, but the main point in the proof is to make the estimates be independent of N. The role of N is now to define the partition Π_α for every allowed pair α.

We have defined the sets X^*, S^*, and Y that do not depend on N and have seen the inequalities (7.9), (7.13) and (8.7)

$$\mathfrak{m}(S^*) \le \frac{7}{y^p}\|f\|_p^p, \quad \mathfrak{m}(X^*) \le \frac{C}{y^p}\|f\|_2^2, \quad \mathfrak{m}(Y) \le \frac{C}{y^p}\|f\|_2^2.$$

Now to define the sets T_N and U_N, we must consider the allowed pairs α and the associated partitions Π_α. Recall from section [4.13] the needed ingredients to define Π_α:

- The function $f \in \mathcal{L}^1(I)$,
- The natural number N,
- A positive real number y,
- The pair α, where $I(\alpha)$ is a smoothing interval of length $|\alpha| \ge 4|I|/2^N$,
- A natural number $j \ge 1$ such that $\||f\||_\alpha < yb_{j-1}$.

We have been given f, N and y, so we consider, for every $j \in \mathbf{N}$, the set of allowed pairs $\alpha \in \mathcal{S}^j$ such that $|\alpha| \ge 4|I|/2^N$ and $\||f\||_\alpha < yb_{j-1}$. For every such pair we obtain the associated partition Π_α. With this partition we can define the functions $\Delta_\alpha(x) = \Delta(\Pi_\alpha, x)$ and $H_\alpha^* f(x) = H_{I(\alpha)}^* E_\alpha f(x)$ and the two sets

$$U_{N,j}(\alpha) = \{x \in I(\alpha) : \Delta_\alpha(x) > C_1 2^m b_{j-1}^{-1/2}\},$$
$$T_{N,j}(\alpha) = \{x \in I(\alpha) : H_\alpha^* f(x) > C_2 2^m y b_{j-1}^{1/2}\},$$
(8.10)

where m is the shift and C_1 and C_2 are constant that we are going to fix.

By the maximal inequalities proved for $\Delta_\alpha(x)$ and $H_\alpha^* f(x)$ (see (5.1) and (5.4)) we obtain

$$\mathfrak{m}(U_{N,j}(\alpha)) \le A|\alpha| \exp\left(-BC_1 2^m b_{j-1}^{-1/2}\right),$$
$$\mathfrak{m}(T_{N,j}(\alpha)) \le A|\alpha| \exp\left(-BC_2 2^m y b_{j-1}^{1/2}\|E_\alpha f\|_\infty^{-1}\right).$$

By (4.18) we have the bound $\|E_\alpha f\|_\infty \le C \sup_k \|f\|_{\beta_k} \le Cyb_{j-1}$. Hence with a proper choice of the constant C_2 in the definition of $T_{N,j}(\alpha)$ we have

$$\mathfrak{m}(T_{N,j}(\alpha)) \le A|\alpha| \exp\left(-BC_1 2^m b_{j-1}^{-1/2}\right).$$

We put

$$U_N = \bigcup_{\alpha,j} U_{N,j}(\alpha), \qquad T_N = \bigcup_{\alpha,j} T_{N,j}(\alpha),$$

where the summation is on all allowed pairs, with $\alpha \in \mathcal{S}^j$, and $\||f\||_\alpha < yb_{j-1}$.

Collecting our results, and selecting adequately C_1, we get

$$\mathfrak{m}(U_N \cup T_N) \le 2\sum_{j=1}^\infty A 2^{-2^{m+8} b_{j-1}^{-1/2}} \sum_{\alpha \in \mathcal{S}^j} |\alpha|.$$

For every $\beta \in \mathcal{R}^{m+j}$ there are eight pairs α such that $\alpha/\beta = \beta$. In fact there are two smoothing intervals $I(\alpha)$ such that $I(\beta)$ is a grandchild of $I(\alpha)$, and there are four possible values of $n(\alpha)$. Thus, taking into account the bound of the length of well situated notes (8.8), we have

$$\leq 32 \sum_{j=1}^{\infty} A 2^{-2^{m+8} b_{j-1}^{-1/2}} \sum_{\beta \in \mathcal{R}^{m+j}, I(\beta) \not\subset X} |\beta| \leq \frac{C \|f\|_2^2}{y^p} \sum_{j=1}^{\infty} b_{m+j}^{-19} 2^{-2^{m+8} b_{j-1}^{-1/2}}.$$

It is easy to see that no matter what the natural number m, there is an absolute constant C such that

$$\mathrm{m}(U_N \cup T_N) \leq \frac{C}{y^p} \|f\|_2^2. \tag{8.11}$$

Another component of the exceptional set is the set D of dyadic points with respect to the interval I.

Finally we define the set V, the last component of E_N. V is defined as the set

$$V = \{x : H_I^* f(x) \geq B y 2^m\}. \tag{8.12}$$

Assume that $f \in \mathcal{L}^p(\mathbf{R})$. By (5.3), $H_I^* f(x) \leq 2 \mathcal{H}^* f(x) + 6 \mathcal{M} f(x)$. The theorems about the maximal Hilbert transform and the Hardy-Littlewood maximal function imply that

$$\|H_I^* f\|_p \leq C \frac{p^2}{p-1} \|f\|_p.$$

Therefore

$$\mathrm{m}(V) \leq \left(\frac{C p^2}{B 2^m (p-1)} \right)^p \frac{\|f\|_p^p}{y^p}.$$

We will see in the following chapter that the selection of m is such that

$$A \frac{p^2}{p-1} \leq 2^m \leq A' \frac{p^2}{p-1}.$$

Therefore choosing conveniently the constant B in (8.12), we get

$$\mathrm{m}(V) \leq \frac{\|f\|_p^p}{y^p}. \tag{8.13}$$

9. Pair Interchange Theorems

9.1 Introduction

This chapter contains the most difficult part of the proof of Carleson's Theorem, that is, how to manage to pass from a Carleson integral $C_\alpha f(x)$ to another where we can apply the basic step. This is accomplished mainly by changing the frequency to obtain $C_{\gamma/\alpha} f(x)$, where γ is an allowed pair with a lesser level than that of the initial Carleson integral. First we will obtain a **nearby allowed pair** β that controls the change of frequency.

This nearby allowed pair will be obtained mainly by hearing the sound of f near α. We will have as our ears Proposition 9.2, which gives a bound of $\|f\|_\alpha$ if f does not sound near α. The structure theorem of the P_u^j will assure us that we will hear a musical sound. We start the chapter choosing the **shift** m. This will allow us to change $y^{p/2}$ for y, but the principal role of the shift m is to classify the quantities that we encounter at two levels of magnitude.

9.2 Choosing the shift m

The measure of the exceptional set E_N must be bounded by $A\|f\|_p^p/y^p$. Of this type is the bound (7.13) of $\mathfrak{m}(S^*)$ and (8.13) of $\mathfrak{m}(V)$. But the bounds we have obtained for $\mathfrak{m}(X^*)$, $\mathfrak{m}(U_N \cup T_N)$ and $\mathfrak{m}(Y)$ in (7.9), (8.11), and (8.7) are of type $A\|f\|_2^2/y^p$. This problem is generated by the use of the Bessel inequality in (7.6). To overcome this difficulty we observe that if f is a measurable function such that $|f| = \chi_A$ is a characteristic function, then $\|f\|_2^2 = \|f\|_p^p$. Later we will have to deal with more general functions. We shall call such a function a **special function**.

There is also another reason for which it is convenient to consider the case of these special function. In fact this will permit us to define the **shift** m, depending only on p, connecting the level of a Carleson integral and y.

The shift m is a natural number, depending only on p. As we have said in the introduction to the definition of allowed pairs S^j, m has a role in the process of selecting a note of f near a not allowed pair. In the following proposition we give to m another role in the proof of the $\mathcal{L}^p$ result.

Proposition 9.1 (Selection of the shift) *Let $1 < p < +\infty$, then there exists a natural number $m = m(p)$ such that if $\alpha \in \mathcal{P}_I$, f is a special function, and $j \in \mathbf{N}$ are such that $b_j y \leq \|\|f\|\|_\alpha$, and $I(\alpha) \not\subset S^*$, then*

$$y^{p/2} \leq b_{m+j}^{-1/4} y, \quad \text{and} \quad b_{m+j} < y. \tag{9.1}$$

Proof. There exists a grandchild J of $I(\alpha)$ such that $\|f\|_{\alpha/J} = \|\|f\|\|_\alpha \geq b_j y$. Since $J \not\subset S$ we have

$$\frac{1}{|J|} \int_J |f|^p \, d\mathbf{m} < y^p.$$

Also from the definition of $\|f\|_{\alpha/J}$ and the fact that $|f|$ is a characteristic function it follows that

$$\|f\|_{\alpha/J} \leq \frac{1}{|J|} \int_J |f| \, d\mathbf{m} = \frac{1}{|J|} \int_J |f|^p \, d\mathbf{m}.$$

Therefore $b_j y < y^p$.

Hence we show that $y^{p/2} \leq b_{m+j}^{-1/4} y$ for every $m \geq m(p)$, if $b_j y < y^p$.

First, we consider the case $1 < p < 2$. For $y \geq 1$ (9.1) is true provided that $m \geq 1$. If $y < 1$, we choose m such that

$$\frac{2^m - 1}{4} > \frac{1 - p/2}{p - 1}.$$

Then we have

$$\frac{2^m - 2^{-j}}{4} > (1 - 2^{-j})\frac{1 - p/2}{p - 1}.$$

We can write that as

$$\frac{1 - 2^{m+j}}{4} < (1 - 2^j)\frac{1 - p/2}{p - 1}.$$

Since $b_j = 2^{1-2^j}$, we deduce that

$$b_{m+j}^{1/4} < b_j^{(1-p/2)/(p-1)} < y^{1-p/2}.$$

This is the first inequality in (9.1). For the second one we consider first the case $1 < p \leq 3/2$, in this case we have $b_{m+j}^{1/4} < y^{1-p/2} \leq y^{1/4}$.

If $3/2 < p < 2$, we observe that $b_j < y^{p-1} < y^{1/2}$, and $b_j^2 > b_{m+j}$.

Now, if $p \geq 2$. For $y \leq 1$ it suffices to take $m \geq 1$, and we have

$$y^{p/2-1} \leq 1 \leq b_{m+j}^{-1/4}, \qquad b_{m+j} < b_j < y^{p-1} \leq y.$$

If $y > 1$, $b_j y \leq 1$. In fact, since $|f|$ is a characteristic function $b_j y \leq \|\|f\|\|_\alpha \leq 1$. We choose m such that

$$\frac{p}{2} - 1 < \frac{2^m - 1}{4}.$$

From which we deduce that

$$(1 - 2^{-j})(p/2 - 1) < (2^m - 2^{-j})/4.$$

Therefore

$$(2^j - 1)(p/2 - 1) < (2^{m+j} - 1)/4.$$

Hence

$$y^{p/2-1} \le b_j^{1-p/2} < b_{m+j}^{-1/4}.$$

□

We shall impose another condition on m, namely that $m \ge m_0$ where m_0 is an absolute constant. This is needed in the proof of Proposition 9.3 below. Roughly speaking, this proposition says that near every Carleson integral $\mathcal{C}_\alpha f(x)$ of level j there is another $\mathcal{C}_\beta f(x)$, with a pair $\beta \in S^j$.

Taking all these restrictions on m into account we can say that there exists a constant $0 < A < +\infty$ such that the shift m can be any natural number satisfying

$$2^m \ge A \frac{p^2}{p-1}.$$

So we have also

$$A' \frac{p^2}{p-1} \ge 2^m,$$

for some absolute constant A'.

9.3 A bound for $\|f\|_\alpha$

The following proposition is essential to obtain, from a lower bound for a local norm: $\|f\|_\alpha > r > 0$, some note of f at the scale of α.

Proposition 9.2 *Let $\alpha = (n, J) \in \mathcal{P}$, $f \in \mathcal{L}^2(J)$ and let*

$$f(t) = \sum_{\beta=(k,J)} a_k e_\beta(t)$$

be its local Fourier expansion. Assume that $|a_k| \le N$ for every k with $|k+n| < M$, where $1 < M < +\infty$, and $0 < N < +\infty$. Then

$$\|f\|_\alpha \le B\left(N \log M + \frac{\|f\|_{\mathcal{L}^2(J)}}{\sqrt{M}}\right), \tag{9.2}$$

where B is an absolute constant.

Proof. By definition

$$\|f\|_\alpha = \sum_{j\in\mathbf{Z}} \frac{c}{1+j^2} \left| \frac{1}{|J|} \int_J f(t) \exp\left(-2\pi i\left(n(\alpha) + \frac{j}{3}\right)\frac{t}{|J|}\right) dt \right|.$$

The integral can be written as

$$\frac{1}{|J|} \int_J f(t) e^{-i\lambda(\alpha)t} \cdot e^{-2\pi ijt/3|J|} \, dt = \sum_{k\in\mathbf{Z}} a'_k \overline{b_k}, \qquad (9.3)$$

where a'_k and b_k are the coefficients of the expansions on $\mathcal{L}^2(J)$

$$f(t)e^{-i\lambda(\alpha)t} = \sum_{\beta=(n,J)} a'_{n(\beta)} e_\beta(t), \qquad e^{2\pi ijt/3|J|} = \sum_{\beta=(n,J)} b_{n(\beta)} e_\beta(t).$$

Hence

$$a'_k = \frac{1}{|J|} \int_J f(t) \exp\left(-2\pi i(n(\alpha) + k)\frac{t}{|J|}\right) dt = a_{n(\alpha)+k}.$$

By hypothesis $|a'_k| \le N$ if $|k| < M$.

If $k + j/3 \ne 0$, the coefficient b_{-k} are bounded by

$$|b_{-k}| \le \left| \frac{\exp\bigl(2\pi i(j/3 + k)\bigr) - 1}{2\pi(j/3 + k)} \right| \le \frac{1}{|j/3 + k|}.$$

In the case $k = -j/3$, $|b_k|$ is bounded by 1.

Now we can bound the integral in (9.3). For $|j| \le M/2$

$$\le \sum_{|k|\le M} |a'_k \overline{b_k}| + \sum_{|k|>M} |a'_k \overline{b_k}|$$

$$\le N + N \sum_{|k|<M, k\ne -j/3} \frac{1}{|j/3 + k|} + \left(\sum_k |a'_k|^2\right)^{1/2} \left(\sum_{|k|\ge M} \left(\frac{1}{|j/3 + k|}\right)^2\right)^{1/2}.$$

Hence there exists a constant C such that

$$\sum_k |a'_k \overline{b_k}| \le C\left(N \log M + \frac{\|f\|_{\mathcal{L}^2(J)}}{\sqrt{M}}\right).$$

For $|j| > M/2$ we use only the Schwarz inequality, and obtain

$$\sum_k |a'_k \overline{b_k}| \le \|f\|_{\mathcal{L}^2(J)}.$$

Therefore $\|f\|_\alpha$ is bounded by

$$C\left(N\log M + \frac{\|f\|_{\mathcal{L}^2(J)}}{\sqrt{M}}\right)\sum_j \frac{c}{1+j^2} + \|f\|_{\mathcal{L}^2(J)}\sum_{|j|>M/2}\frac{1}{1+j^2}$$

$$\leq B\left(N\log M + \frac{\|f\|_{\mathcal{L}^2(J)}}{\sqrt{M}}\right).$$

$\square$

9.4 Selecting an allowed pair

Every grandson α/K of a not allowed pair $\alpha \notin \mathcal{S}^j$ is not well situated ($\alpha/K \notin \mathcal{R}^{m+j}$). Then, we have arranged things so that the sound f in the scale of α/K is a rest or a pure note. If we also assume that we are at level j ($b_j y \leq \||f\||_\alpha < b_{j-1}y$), and have selected adequately the shift m, then f definitely sounds, so by the structure theorem $P^{m+j}_{u(\delta/K)}$ must be a pure note. All pure notes of f are well situated, so a grandfather β of this note will be an allowed pair nearby to α. If we are considering this β as a step in our process, we must bound the difference $|\mathcal{C}_{\beta/\alpha}f(x) - \mathcal{C}_\alpha f(x)|$, for this will be convenient to give a bound for $|n(\beta/\alpha) - n(\alpha)|$.

In this paragraph we are undertaking the process of selecting this **nearby allowed pair** β.

Proposition 9.3 *Let f be a special function, and $\alpha \in \mathcal{P}_I$ such that $b_j y \leq \||f\||_\alpha < b_{j-1}y$, $\alpha \notin \mathcal{S}^j$, and let $x \in I/2$ be a point such that $x \in I(\alpha)/2$ and $x \notin D \cup S^* \cup X^* \cup Y$. Then there exists $\beta \in \mathcal{S}^j$, with $I(\beta) \supset I(\alpha)$, $x \in I(\beta)/2$, and such that*

$$\left|n(\alpha) - n(\beta/\alpha)\right| < \frac{A_0}{b_j}, \tag{9.4}$$

and for every γ with $I(\gamma) = I(\alpha)$, and $|n(\alpha) - n(\gamma)| \leq 2A_0 b_j^{-3/2}$ we have

$$\left|\mathcal{C}_\gamma f(x) - \mathcal{C}_\alpha f(x)\right| \leq B\left(\||f\||_{\beta/\alpha} + b_j y\right). \tag{9.5}$$

Remark. In such a situation we shall call β a nearby allowed pair to α. Its function is to control the changes of frequency.

Proof. There exists a grandchild K of $I(\alpha)$ such that $\|f\|_{\alpha/K} = \||f\||_\alpha$. For this K, as for every other grandchild of $I(\alpha)$, we have $\alpha/K \notin \mathcal{R}^{m+j}$. We hope to find a β starting with some term of the function $P^{m+j}_u(t)$, where $u = u(K)$.

A general remark about the proof: Thanks to the shift m, we have quantities at two levels yb_j and yb_{m+j} and always $yb_j \gg yb_{m+j}$.

We divide the proof in four steps.

First step: To obtain $\delta \in \mathcal{Q}^{m+j}$, **with** $I(\delta) \supset K$

The proof starts applying Proposition 9.2 to obtain, for every $k \in \mathbf{Z}$, a bound for $\|f - P_u^{m+j}\|_{(k,K)}$.

By construction of $P_u^{m+j}(t)$ the local Fourier coefficients of $f - P_u^{m+j}(t)$ are less than $b_{m+j} y^{p/2}$.

Also

$$\|f - P_u^{m+j}\|_{\mathcal{L}^2(K)} \le \|f\|_{\mathcal{L}^2(K)} + \|P_u^{m+j}\|_{\mathcal{L}^2(K)}. \tag{9.6}$$

Since $I(\alpha) \not\subset S^*$, $K \not\subset S$ and $\|f\|_{\mathcal{L}^2(K)}^2 = \|f\|_{\mathcal{L}^p(K)}^p < y^p$. Furthermore, since $I(\alpha) \not\subset X^*$, $K \not\subset X$ and by (7.8), $|P_u^{m+j}(x)| \le y^{p/2} b_{m+j}^{-2}$. Collecting all this we obtain

$$\|f - P_u^{m+j}\|_{\mathcal{L}^2(K)} \le y^{p/2} + b_{m+j}^{-2} y^{p/2} \le 2 b_{m+j}^{-2} y^{p/2}.$$

We are in position to apply Proposition (9.2) with $N = b_{m+j} y^{p/2}$, $M = b_{m+j}^{-8}$ and the bound obtained for $\|f - P_u^{m+j}\|_{\mathcal{L}^2(K)}$. We obtain that for every $k \in \mathbf{Z}$

$$\|f - P_u^{m+j}\|_{(k,K)} \le B\left(y^{p/2} b_{m+j} \log(b_{m+j}^{-8}) + \frac{2 b_{m+j}^{-2} y^{p/2}}{b_{m+j}^{-4}}\right)$$

$$\le B y^{p/2}\left(b_{m+j} \log(b_{m+j}^{-8}) + 2 b_{m+j}^2\right) \le 3 B y^{p/2} b_{m+j}^{3/2}. \tag{9.7}$$

The last inequality requires that we take m big enough. In fact $m \ge 5$ suffices.

Now we have a lower bound of $\|P_u^{m+j}\|_{\alpha/K}$. This will imply that there are terms in this function.

$$\|P_u^{m+j}\|_{\alpha/K} \ge \|f\|_{\alpha/K} - \|f - P_u^{m+j}\|_{\alpha/K} \ge y b_j - 3 B y^{p/2} b_{m+j}^{3/4}$$

It is here were we are forced to choose the shift so that $y^{p/2} \le b_{m+j}^{-1/4} y$ or something similar.

Since f is a special function, we are able to choose the shift so that $y^{p/2} \le y b_{m+j}^{-1/4}$, therefore

$$\|P_u^{m+j}\|_{\alpha/K} \ge y b_j - 3 B y b_{m+j}^{1/2} > \frac{1}{2} y b_j. \tag{9.8}$$

We assume here that m has been chosen in order to satisfy

$$3 B b_{m+j}^{1/2} < b_j/2.$$

Given that $b_j = 2^{1-2^j}$, it is easy to see that this condition is equivalent to $m \ge m_0$ for some absolute constant m_0.

Note that the same calculation proves that for every grandchild $L = I_v$ of $I(\alpha)$ we have

$$\|f - P_v^{m+j}\|_{(k,L)} \le 3By b_{m+j}^{1/2}. \tag{9.9}$$

Since $\alpha/K \notin \mathcal{R}^{m+j}$ and $K \not\subset X$, the lemma on the structure of P_u^{m+j} (lemma 8.1) says that

$$P_u^{m+j}(t) = \rho e^{i\lambda(\delta)t} + P_0(t) + P_1(t), \qquad t \in K = I_u \tag{9.10}$$

where

$$|\rho| \le b_{m+j}^{-2} y^{p/2}, \qquad |P_0(t)| \le b_{m+j}^8 y^{p/2}, \tag{9.11}$$

and all the terms $a(\gamma)e^{i\lambda(\gamma)t}$ of $P_1(t)$, satisfy $|n(\alpha/K) - n(\gamma/K)| \ge b_{m+j}^{-10}$.

What we want to prove is that $\rho \ne 0$ so that $\delta \in \mathcal{Q}^{m+j}$ is such that $I(\delta) \supset K$. Then we can choose β to be a grandfather of δ. But $\rho = 0$ would imply that $\|P_u^{m+j}\|_{\alpha/K}$ is small and this will be in contradiction with (9.8). So we try to obtain an upper bound of $\|P_u^{m+j}\|_{\alpha/K}$, related to the decomposition (9.10) and to compare it with (9.8).

We have given adequate bounds for the local norm of exponentials in Proposition 4.4. Hence taking (9.10) into account we have

$$\|P_u^{m+j}\|_{\alpha/K} \le \frac{C|\rho|}{|n(\delta/K) - n(\alpha/K)|} + b_{m+j}^8 y^{p/2} + C b_{m+j}^{10} \sum_\gamma{}'' |a(\gamma)|.$$

If $n(\delta/K) = n(\alpha/K)$ the first term is reduced to $|\rho|$.

Since $K \not\subset X$, we have a bound for the last sum (see (7.8))

$$\|P_u^{m+j}\|_{\alpha/K} \le \frac{C|\rho|}{|n(\delta/K) - n(\alpha/K)|} + b_{m+j}^8 y^{p/2} + C b_{m+j}^8 y^{p/2}.$$

We conclude from (9.8) that

$$\frac{1}{2} y b_j < \|P_u^{m+j}\|_{\alpha/K} \le \frac{C|\rho|}{|n(\delta/K) - n(\alpha/K)|} + 2 b_{m+j}^7 y.$$

We assume m is such that, for every j, $8 b_{m+j}^7 < b_j$, and we obtain

$$y b_j < \frac{4C|\rho|}{|n(\delta/K) - n(\alpha/K)|}. \tag{9.12}$$

Therefore $|\rho| \ne 0$. This gives us a term $\delta \in \mathcal{Q}^{m+j}$ and such that $I(\delta) \supset K$.

Second step: Bound for $|n(\delta/K) - n(\alpha/K)|$.

Now we must prove that the pitch of δ is comparable to that of α. That is we must bound $|n(\delta/K) - n(\alpha/K)|$, so we assume it is not null.

From the proof of the structure theorem (of P_u^j) we know that we can take as δ any pair $\in \mathcal{Q}^{m+j}$ such that $I(\delta) \supset K$ and such that there holds the inequality $|n(\delta/K) - n(\alpha/K)| < b_{m+j}^{-10}$. But our hypotheses $b_j y \le \|\|f\|\|_\alpha <$

$b_{j-1}y$ and $\alpha \notin S^j$ are very strong. In fact, we can prove that every one of this δ's satisfies $|n(\delta/K) - n(\alpha/K)| < Cb_j^{-1}$. To this end we must prove $|\rho| \leq Cy$ (cf. (9.12)). A first approximation is obtained from (9.12) and the known estimate $|\rho| \leq b_{m+j}^{-2} y^{p/2}$

$$|n(\delta/K) - n(\alpha/K)| \leq \frac{4Cb_{m+j}^{-2} y^{p/2}}{yb_j} \leq 4Cb_{m+j}^{-4} < b_{m+j}^{-5}. \tag{9.13}$$

Now we reverse the inequalities from the previous step.

The local norm $\|P_u^{m+j}\|_{(k,K)}$ has a maximum near $k = n(\delta/K)$. Hence for $\delta/K = (k, K)$, Proposition 4.5 gives us

$$\|P_u^{m+j}\|_{\delta/K} \geq B|\rho| - b_{m+j}^8 y^{p/2} - \sum_{\gamma}'' \frac{C|a(\gamma)|}{|n(\delta/K) - n(\gamma/K)|}.$$

And we have

$$\sum_{\gamma}'' \frac{C|a(\gamma)|}{|n(\delta/K) - n(\gamma/K)|} \leq$$

$$\leq \sum_{\gamma}'' \frac{C|a(\gamma)|}{|n(\gamma/K) - n(\alpha/K)| - |n(\delta/K) - n(\alpha/K)|}.$$

Hence, by the sense of $\sum''$ and (9.13), this sum is bounded by

$$\sum_{\gamma}'' \frac{C|a(\gamma)|}{b_{m+j}^{-10} - b_{m+j}^{-5}} \leq \frac{Cy^{p/2}b_{m+j}^{-2}}{b_{m+j}^{-10} - b_{m+j}^{-5}}.$$

Since $m \geq 2$ we obtain $2b_{m+j}^{-5} < b_{m+j}^{-10}$, so that finally we arrive at

$$\|P_u^{m+j}\|_{\delta/K} \geq B|\rho| - b_{m+j}^8 y^{p/2} - 2Cy^{p/2}b_{m+j}^8 > B|\rho| - yb_{m+j}^6.$$

But, on the other hand, by (9.7),

$$\|P_u^{m+j}\|_{\delta/K} \leq \|f\|_{\delta/K} + \|f - P_u^{m+j}\|_{\delta/K} \leq \|f\|_{\delta/K} + 3Byb_{m+j}^{1/2}.$$

Therefore

$$|\rho| \leq C(\|f\|_{\delta/K} + yb_{m+j}^{1/2}). \tag{9.14}$$

Now $\|f\|_{\delta/K} \leq \|f\|_{\mathcal{L}^1(K)} \leq \|f\|_{\mathcal{L}^p(K)} < y$ and we see at once that

$$|\rho| \leq 2Cy,$$

which with (9.12) establishes that

$$|n(\delta/K) - n(\alpha/K)| < \frac{C}{b_j}.$$

Third step: Definition of β

We are in position to apply Lemma 8.2, to prove that the functions P_v^{m+j} are the same for every grandson of α. Namely α is a pair such that $\alpha/L \notin \mathcal{R}^{m+j}$ for every grandchild L of $I(\alpha)$, $I(\alpha) \not\subset Y$ by hypothesis, and there is a term $\delta \in \mathcal{Q}^{m+j}$ of P_u^{m+j} such that

$$|n(\delta/K) - n(\alpha/K)| < \frac{C}{b_j} < b_{m+j}^{-9}.$$

(This is another restriction on m of the type $m \geq m_0$). We deduce that in fact $I(\delta) \supset I(\alpha)$ and that the four functions P_v^{m+j}, corresponding to the four grandsons I_v of $I(\alpha)$, coincide.

Since $\delta \in \mathcal{Q}^{m+j}$ we know that δ is a dyadic interval with $|\delta| \leq |I|/4$. Then as $x \in I/2$ is not a dyadic point there is only one smoothing interval $I(\beta)$ of length $4|\delta|$ and such that $x \in I(\beta)/2$. We define β as the pair with $n(\beta) = 4n(\delta)$, and this $I(\beta)$. It is easy to see that with this definition $I(\delta)$ is a grandson of $I(\beta)$ and $\beta/I(\delta) = \delta$. Furthermore $I(\beta) \not\subset X^*$ since $I(\beta) \supset I(\delta) \supset I(\alpha)$ and $I(\alpha) \not\subset X^*$. Hence to prove $\beta \in \mathcal{S}^j$ we only have to prove that $\delta \in \mathcal{R}^{m+j}$.

We know that $\delta \in \mathcal{Q}^{m+j}$. But condition $A(\mathcal{R}^j)$ implies that every $\alpha \in \mathcal{Q}^j$ satisfies also $\alpha \in \mathcal{R}^j$ (take β and γ equal to α in condition $A(\mathcal{R}^j)$). Hence $\delta \in \mathcal{R}^{m+j}$.

We also have that

$$|n(\alpha) - n(\beta/\alpha)| = \left| n(\alpha) - \left\lfloor n(\beta)\frac{|\alpha|}{|\beta|} \right\rfloor \right| = \left| 4n(\alpha/K) + r - \left\lfloor 4n(\delta)\frac{|K|}{|\delta|} \right\rfloor \right|$$

$$\leq 4\frac{C}{b_j} + 8 \leq \frac{A_0}{b_j}.$$

This proves (9.4).

Fourth step: Bound for $|\mathcal{C}_\gamma f(x) - \mathcal{C}_\alpha f(x)|$.

Let γ be a pair such that $I(\gamma) = I(\alpha)$ and $|n(\alpha) - n(\gamma)| \leq 2A_0 b_j^{-3/2}$. Let $P = P_u^{m+j}$ and recall that we have proved that it coincides with every P_v^{m+j} if $v = v(L)$ for some grandchild L of $I(\alpha)$. Also note that (9.10) and the inequalities in (9.11) are valid now for every $t \in I(\alpha)$.

First we have

$$|\mathcal{C}_\gamma f(x) - \mathcal{C}_\alpha f(x)| \leq |\mathcal{C}_\gamma (f - P)(x) - \mathcal{C}_\alpha (f - P)(x)| + |\mathcal{C}_\gamma P(x) - \mathcal{C}_\alpha P(x)|.$$

By a changing of frequency, by the bounds of $\|f - P\|_{(k,L)}$ obtained at step 1 (cf. (9.9)), and by the comparison between the two norms $\|f - P\|_\alpha$ and $\|\|f - P\|\|_\alpha$ (proposition (4.12)) we get

$$|\mathcal{C}_\gamma (f - P)(x) - \mathcal{C}_\alpha (f - P)(x)| \leq C(2A_0 b_j^{-3/2})^3 3By b_{m+j}^{1/2} \leq b_j y,$$

(by the restriction $m \geq m_0$).

Second, by the structure of P

$$|\mathcal{C}_\gamma P(x) - \mathcal{C}_\alpha P(x)| \le |\rho| \cdot |\mathcal{C}_\gamma(e^{i\lambda(\delta)t})(x) - \mathcal{C}_\alpha(e^{i\lambda(\delta)t})(x)|$$
$$+ |\mathcal{C}_\gamma(P_0)(x) - \mathcal{C}_\alpha(P_0)(x)| + |\mathcal{C}_\gamma(P_1)(x) - \mathcal{C}_\alpha(P_1)(x)|.$$

The Carleson integral of an exponential is bounded (cf. Lemma 4.1). So for the first term we have

$$|\rho| \cdot |\mathcal{C}_\gamma(e^{i\lambda(\delta)t})(x) - \mathcal{C}_\alpha(e^{i\lambda(\delta)t})(x)| \le 2A|\rho|.$$

In the second term we have, one more time, a change of frequency. Hence

$$|\mathcal{C}_\gamma(P_0)(x) - \mathcal{C}_\alpha(P_0)(x)| \le B(2A_0 b_j^{-3/2})^3 \|P_0\|_\alpha \le A b_j^{-9/2} b_{m+j}^7 y < b_j y.$$

(Again by the assumption $m \ge m_0$).

The third term is bounded in the following way:

$$|\mathcal{C}_\gamma(P_1)(x) - \mathcal{C}_\alpha(P_1)(x)| \le \sideset{}{''}\sum_\eta |a(\eta)| \cdot |\mathcal{C}_\gamma(e^{i\lambda(\eta)t}) - \mathcal{C}_\alpha(e^{i\lambda(\eta)t})|.$$

By another change of frequency, we get

$$|\mathcal{C}_\gamma(P_1)(x) - \mathcal{C}_\alpha(P_1)(x)|$$
$$\le C \sideset{}{''}\sum_\eta |a(\eta)| b_j^{-9/2} \|e^{i\lambda(\eta)t}\|_\alpha \le C b_j^{-9/2} \sideset{}{''}\sum_\eta \frac{|a(\eta)|}{\left| \left\lfloor \frac{n(\eta)}{|\eta|}|\alpha| \right\rfloor - n(\alpha) \right|}$$
$$\le C b_j^{-9/2} \sideset{}{''}\sum_\eta \frac{|a(\eta)|}{4|n(\eta/K) - n(\alpha/K)| - 8} \le C' b_j^{-9/2} b_{m+j}^{10} y^{p/2} b_{m+j}^{-2}$$
$$\le C b_j^{-9/2} b_{m+j}^7 y \le b_j y.$$

Collecting all these inequalities we have proved that

$$|\mathcal{C}_\gamma f(x) - \mathcal{C}_\alpha f(x)| \le 2A|\rho| + 3b_j y \le C(|\rho| + b_j y),$$

and by (9.14)

$$|\mathcal{C}_\gamma f(x) - \mathcal{C}_\alpha f(x)| \le B(\|f\|_{\delta/K} + b_j y) \le B(\|\|f\|\|_{\beta/\alpha} + b_j y),$$

since $(\beta/\alpha)/K = \delta/K$.   □

Given a Carleson integral $\mathcal{C}_\alpha f(x) \ne 0$ with $x \notin S^*$, it has a level $j \in \mathbf{N}$ (i. e. $b_j y \le \|\|f\|\|_\alpha < b_{j-1} y$). Moreover, if $\alpha \in \mathcal{S}^j$ we can apply the basic step. If, on the other hand, $\alpha \notin \mathcal{S}^j$, by the previous theorem, there exists what we call a nearby allowed pair $\beta \in \mathcal{S}^j$ with $x \in I(\beta)/2$, $I(\alpha) \subset I(\beta)$, that controls the changes of frequencies, so that if $|n(\gamma/\alpha) - n(\alpha)| < 2A_0 b_j^{-3/2}$, we have

$$|\mathcal{C}_{\gamma/\alpha} f(x) - \mathcal{C}_\alpha f(x)| \le B(\|\|f\|\|_{\beta/\alpha} + b_j y).$$

In particular, this is true for $\gamma = \beta$.

Since β is an allowed pair $(\beta \in \mathcal{S}^j)$ and

$$|\mathcal{C}_{\beta/\alpha}f(x) - \mathcal{C}_\alpha f(x)| \leq B(|||f|||_{\beta/\alpha} + b_j y)$$

we can think to apply the basic step to the Carleson integral $\mathcal{C}_{\beta/\alpha}f(x)$. This can be problematic for two reasons: (a) We have not a reasonable bound for $|||f|||_{\beta/\alpha}$. (b) We shall arrive to a Carleson integral $\mathcal{C}_{\beta/I(x)}f(x)$. There is not guarantee that $I(x) \subsetneq I(\alpha)$. This situation is inaceptable. What we need is a more simple Carleson integral, not one with more cycles. Therefore we have not obtained our objective. What is the new level k with $b_k y \leq |||f|||_{\beta/\alpha} < b_{k-1}y$?, is $\beta/\alpha \in \mathcal{S}^k$? These questions have not a unique answer.

Since the best bound of $|||f|||_{\beta/\alpha}$ is $b_{k-1}y$, the next time we must apply the basic step at some level $\leq k$, and it is essential that $k \leq j$, because we started at level j.

Therefore what we need is a pair ξ, and a level ℓ such that:

(a) We are in a good position to apply the basic step, $I(\xi) \supset I(\alpha)$,

$$x \in I(\xi)/2, \qquad \xi \in \mathcal{S}^\ell, \quad \text{and} \quad |||f|||_\xi < b_{\ell-1}y.$$

(b) We are at an adequate level

$$|||f|||_{\beta/\alpha} < b_{\ell-1}y, \text{ with } \ell < j.$$

(c) We have a controlled change of frequency

$$|n(\xi/\alpha) - n(\alpha)| < 2A_0 b_j^{-3/2}.$$

(d) Finally, to be able to apply the basic step, $I(\alpha)$ must be a union of sets of the partition Π_ξ corresponding to this pair ξ.

This pair ξ is obtained by picking, from the set of (μ, ℓ) with some of these properties, one with the minimum level ℓ. Next we prove that it satisfies all our conditions.

Proposition 9.4 *Assume that f is a special function, $x \in I/2$ and let $\alpha \in \mathcal{P}_I$, such that $b_j y \leq |||f|||_\alpha < b_{j-1}y$, $\alpha \notin \mathcal{S}^j$ and $x \in I(\alpha)/2$ such that $x \notin D \cup S^* \cup X^* \cup Y$. We also assume that $0 \leq n(\alpha) < 2^N$, and $|\alpha| > 4|I|/2^N$.*

Then there exists $\xi \in \mathcal{S}^k$, with $1 \leq k \leq j$ such that $I(\xi) \supset I(\alpha)$, $x \in I(\xi)/2$, and also with β being a nearby allowed pair to α, it is satisfied that

$$|||f|||_{\beta/\alpha} \leq b_{k-1}y, \qquad |||f|||_\xi < b_{k-1}y, \qquad |n(\xi/\alpha) - n(\alpha)| \leq \frac{2A_0}{b_j}. \qquad (9.15)$$

Furthermore, if Π_ξ is the partition determined on $I(\xi)$ and $I(x)$ is the central interval, then $I(x) \subsetneq I(\alpha)$, and $I(\alpha) \setminus I(x)$ is a union of intervals of Π_ξ.

Proof. Let Σ be the set of pairs (μ, ℓ), where $\mu \in \mathcal{P}_I$ and $\ell \in \mathbf{N}$ satisfying the conditions:

(i) $I(\mu) \supset I(\alpha)$ and $x \in I(\mu)/2$.
(ii) $\|\|f\|\|_{\beta/\alpha} < b_{\ell-1}y$, and $\ell \leq j$.
(iii) $\left|n(\mu/\alpha) - n(\alpha)\right| \leq A_0 \sum_{i=\ell}^{j} b_i^{-1}$.
(iv) $\mu \in \mathcal{S}^\ell$.

Where in (iii) the constant A_0 is the same as that appearing in proposition 9.3.

The proof will be divided into 3 steps.

Step 1: Σ is nonempty.

If $\|\|f\|\|_{\beta/\alpha} < b_{j-1}y$, then $(\beta, j) \in \Sigma$ which proves our claim.

If $\|\|f\|\|_{\beta/\alpha} \geq b_{j-1}y$ there exists k, with $1 \leq k \leq j-1$ such that $b_k y \leq \|\|f\|\|_{\beta/\alpha} < b_{k-1}y$. (Observe that, since $I(\beta/\alpha) \not\subset S^*$, we know that $\|\|f\|\|_{\beta/\alpha} < y$, and $\|\|f\|\|_{\beta/\alpha} > 0$ since $\|\|f\|\|_\alpha \geq b_j y > 0$).

Now, if $\beta/\alpha \in \mathcal{S}^k$, then $(\beta/\alpha, k) \in \Sigma$.

If on the other hand $\beta/\alpha \notin \mathcal{S}^k$, we obtain an allowed pair β' nearby to β/α (proposition 9.3); hence $\beta' \in \mathcal{S}^k$, with $I(\beta') \supset I(\beta/\alpha) = I(\alpha)$, $x \in I(\beta')/2$, and

$$\left|n(\beta/\alpha) - n(\beta'/\alpha)\right| \leq \frac{A_0}{b_k}.$$

Then $(\beta', k) \in \Sigma$. We see that condition (iii) in the definition of Σ is satisfied since

$$\left|n(\beta'/\alpha) - n(\alpha)\right| \leq \left|n(\beta'/\alpha) - n(\beta/\alpha)\right| + \left|n(\beta/\alpha) - n(\alpha)\right|$$

$$\leq \frac{A_0}{b_k} + \frac{A_0}{b_j} \leq A_0 \sum_{i=k}^{j} b_i^{-1}.$$

The other conditions are easily verified.

Step 2: Selection of $(\xi, k) \in \Sigma$ and the proof of $\|\|f\|\|_\xi < b_{k-1}y$.

We pick a $(\xi, k) \in \Sigma$ with a minimum k. We are going to prove that ξ and k satisfy the theorem.

If it were true that $\|\|f\|\|_\xi \geq b_{k-1}y$ there would be an ℓ, with $1 \leq \ell < k$ and $b_\ell y \leq \|\|f\|\|_\xi < b_{\ell-1}y$.

If $\xi \in \mathcal{S}^\ell$, $(\xi, \ell) \in \Sigma$ in contradiction to the selection of (ξ, k).

Therefore $\xi \notin \mathcal{S}^\ell$. Then (by Proposition 9.3) there exists an allowed pair $\beta'' \in \mathcal{S}^\ell$ nearby to ξ. Then $I(\beta'') \supset I(\xi)$, $x \in I(\beta'')/2$, and $\left|n(\xi) - n(\beta''/\xi)\right| < A_0/b_\ell$.

Since $(\xi, k) \in \Sigma$, we have $\left|n(\xi/\alpha) - n(\alpha)\right| \leq A_0 \sum_{i=k}^{j} b_i^{-1}$, $I(\xi) \supset I(\alpha)$, and $\|\|f\|\|_{\beta/\alpha} < b_{k-1}y$.

We can check now that $(\beta'', \ell) \in \Sigma$. Condition (i) is clear; (ii) follows from $\|\|f\|\|_{\beta/\alpha} < b_{k-1}y < b_{\ell-1}y$.

To deduce (iii), observe that

$$\left|n(\beta''/\alpha) - n(\alpha)\right| \leq \left|n(\beta''/\alpha) - n(\xi/\alpha)\right| + \left|n(\xi/\alpha) - n(\alpha)\right|$$

$$\leq \left|n(\beta''/\alpha) - n(\xi/\alpha)\right| + A_0 \sum_{i=k}^{j} b_i^{-1}.$$

Now $\left|n(\beta''/\alpha) - n(\xi/\alpha)\right| \leq \left|n(\beta''/\xi) - n(\xi)\right|$. This is a particular case of the inequality

$$\left|\left\lfloor \frac{N}{2^{r+s}} \right\rfloor - \left\lfloor \frac{M}{2^r} \right\rfloor\right| \leq \left|\left\lfloor \frac{N}{2^s} \right\rfloor - M\right|,$$

valid when M, N, r and s are nonnegative integers. This can be easily proved using the binary expansion of natural numbers.

Therefore we have

$$\left|n(\beta''/\alpha) - n(\alpha)\right| \leq \frac{A_0}{b_\ell} + A_0 \sum_{i=k}^{j} b_i^{-1} \leq A_0 \sum_{i=\ell}^{j} b_i^{-1}.$$

Since $\ell < k$, that $(\beta'', \ell) \in \Sigma$ is in contradiction to the definition of (ξ, k). This contradiction proves the inequality $\|\|f\|\|_\xi < b_{k-1}y$, and finishes step 2.

Step 3: Claims on the partition Π_ξ.

Now let Π_ξ be the partition of $I(\xi)$ obtained with the given N, y, and the inequality $\|\|f\|\|_\xi < b_{k-1}y$. Let also $I(x)$ denote the central interval determined by x and Π_ξ.

$I(x)$ and $I(\alpha)$ are two smoothing intervals which contain x in their middle halves. By Proposition 4.6 it follows that $I(\alpha) \subset I(x)$ or $I(x) \subsetneq I(\alpha)$.

The first hypothesis leads to a contradiction. In fact $I(x) \supset I(\alpha)$ and $|\alpha| > 4|I|/2^N$. By construction, one of the two halves of $I(x)$, say L, is a member of the partition Π_ξ. Since $|L| > 2|I|/2^N$, K, one of the two sons of L, is such that $\|f\|_{\xi/K} \geq b_{k-1}y$. Hence there exists ℓ with $1 \leq \ell < k$ such that $b_\ell y \leq \|\|f\|\|_{\xi/I(x)} < b_{\ell-1}y$.

If $\xi/I(x) \in \mathcal{S}^\ell$, then $(\xi/I(x), \ell) \in \Sigma$. This is in contradiction to the selection of (ξ, k) in Σ.

If $\xi/I(x) \notin \mathcal{S}^\ell$, Proposition 9.3 gives a nearby allowed pair $\gamma \in \mathcal{S}^\ell$. As in the second step we can prove that $(\gamma, \ell) \in \Sigma$. This is also contradictory.

We have proved $I(x) \subsetneq I(\alpha) \subset I(\xi)$. By the last condition of Proposition 4.7, in this case $I(\alpha)$ and $I(x)$ are unions of intervals of the partition Π_ξ. $\quad\square$

10. All together

10.1 Introduction

In this chapter we prove the weak inequality from which all the results about the Carleson maximal operator can be obtained by standard methods.

In Proposition 10.1, we apply the results of the previous chapter to prove that given a Carleson integral we can either bound it at its level or we can obtain another Carleson integral for which we can bound the difference and such that the second Carleson integral has a lesser level.

Theorem 10.2 is the principal result of Carleson-Hunt. After proving this, it will only remain to show that all the pieces of the jigsaw fit together.

In particular, we have to adjust the absolute constant θ so that before we arrive at a Carleson integral with $|\alpha| \leq 4|I|/2^N$ (for which we can not apply the basic step or Proposition 9.4) we arrive at one for which $n(\alpha) = 0$.

10.2 End of the proof

The following proposition is a formulation of the basic step of the proof. We can summarize it by saying that a Carleson integral of level j is small with respect to its level or can be approximated (in relation to its level) by another Carleson integral of level $j' < j$. In this chapter, if $\|f\|_\alpha = 0$, we shall say that the level of the Carleson integral $C_\alpha f(x)$ is ∞. Thus the level of a Carleson integral is always defined and is a natural number or infinity. Throughout this chapter f is a special function, $|f| = \chi_A$. Also we denote by E_N the exceptional set defined in section [8.5]. Therefore $\mathrm{m}(E_N) \leq C\|f\|_p^p/y^p$.

Proposition 10.1 Let $C_\alpha f(x)$ be a Carleson integral of level j, such that $x \notin E_N$, and $|\alpha| > 4|I|/2^N$. Then there exists a natural number $1 \leq k \leq j$ satisfying one of the two following conditions:

(A) $|C_\alpha f(x)| \leq C 2^m y b_{k-1}^{1/2}$.

(B) There exists $\delta \in \mathcal{P}_I$, such that $I(\delta) \subsetneq I(\alpha)$, $x \in I(\delta)/2$, $\lambda(\delta) \leq (1 + b_j^{1/2})\lambda(\alpha)$, and either

$$|\delta| \leq 4|I|/2^N$$

or the level of $C_\delta f(x)$ is $j' < k$, and

$$\left| C_\delta f(x) - C_\alpha f(x) \right| \le C2^m y b_{k-1}^{1/2}. \tag{10.1}$$

Proof. If the level of $C_\alpha f(x)$ is ∞, then $C_\alpha f(x) = 0$ and condition (A) is satisfied with $k = 1$. Thus we assume that $j \in \mathbf{N}$.

Because j is the level of $C_\alpha f(x)$ we have $y b_j \le \|\|f\|\|_\alpha < y b_{j-1}$. We divide the proof into two cases.

First case. Assume that α is an allowed pair $\alpha \in \mathcal{S}^j$. Given N and $\|\|f\|\|_\alpha < y b_{j-1}$, we can use the procedure to find a partition Π_α of $I(\alpha)$ and a smoothing interval $I(x)$ as in Proposition 4.7.

Set $\delta = \alpha/I(x)$. Then, δ satisfies the condition of the theorem, in particular

$$\lambda(\delta) = 2\pi \frac{n(\delta)}{|\delta|} \le 2\pi\, n(\alpha) \frac{|\delta|}{|\alpha|} \frac{1}{|\delta|} = \lambda(\alpha).$$

If $|\delta| > 4|I|/2^N$, we can apply the basic step, with $\xi = \alpha$ and $J = I(\alpha)$ and obtain the bound

$$|C_\alpha f(x) - C_\delta f(x)| \le C y b_{j-1} + 2H_\alpha^* f(x) + D y b_{j-1}\, \Delta_\alpha(x).$$

Since $\alpha \in \mathcal{S}^j$ and $x \notin T_N(\alpha) \cup U_N(\alpha)$ we have by the definition of the sets $T_{N,j}(\alpha)$ and $U_{N,j}(\alpha)$

$$H_\alpha^* f(x) \le C_2 2^m y b_{j-1}^{1/2}, \qquad \Delta_\alpha(x) \le C_1 2^m b_{j-1}^{-1/2}.$$

And we get

$$|C_\alpha f(x) - C_\delta f(x)| \le C2^m y b_{j-1}^{1/2}.$$

In this case we take $k = j$ and condition (B) is satisfied. Since $|\delta| > 4|I|/2^N$, the construction of $I(x)$ also implies that the level j' of $C_\delta f(x)$ is less than $k = j$, and $I(x) \subsetneq I(\alpha)$.

Second case. Assume that $\alpha \notin \mathcal{S}^j$. We apply the machinery of the last chapter: We select a nearby allowed pair β and apply Proposition 9.4, finding a level k, with $1 \le k \le j$ and a pair $\xi \in \mathcal{S}^k$ such that $I(\xi) \supset I(\alpha)$, $x \in I(\xi)/2$, and $\|\|f\|\|_{\beta/\alpha} \le y b_{k-1}$.

Let also Π_ξ and $I(x)$ be the partition and central interval arising from ξ and the inequality $\|\|f\|\|_\xi < y b_{k-1}$.

First subcase. If $n(\alpha) \le 2A_0/b_j^{3/2}$, since the nearby allowed pair β controls the changes of frequency, we have

$$\left| C_\gamma f(x) - C_\alpha f(x) \right| \le B(\|f\|_{\beta/\alpha} + b_j y) \le C y b_{k-1},$$

where $\gamma = (0, I(\alpha))$. Now as $x \notin V$

$$\left| \mathcal{C}_\gamma f(x) \right| = \left| \text{p.v.} \int_{I(\alpha)} \frac{f(t)}{x - t} \, dt \right| \leq B 2^m y.$$

Thus, in this subcase, we put $k = 1$ (changing its previous value) and obtain
$\left| \mathcal{C}_\alpha f(x) \right| \leq C 2^m y b_{k-1}^{1/2} = C 2^m y.$

Second subcase. Now we assume $n(\alpha) > 2A_0/b_j^{3/2}$. Take $\delta = \xi/I(x)$.

(A) Proof of $\lambda(\delta) \leq (1 + b_j^{1/2})\lambda(\alpha)$.
Observe that by definition and by the construction of ξ

$$n(\delta) = \left\lfloor n(\xi) \frac{|I(x)|}{|\xi|} \right\rfloor, \qquad \left| n(\xi/\alpha) - n(\alpha) \right| \leq \frac{2A_0}{b_j}.$$

Put $n' = n(\xi/\alpha)$. Since $I(\xi) \supset I(\alpha) \supsetneq I(\delta) = I(x)$, we get

$$n(\delta) = n(\xi/I(x)) = n((\xi/\alpha)/I(x)) = \left\lfloor n' \frac{|\delta|}{|\alpha|} \right\rfloor.$$

Therefore

$$\frac{n(\delta)}{|\delta|} = \frac{1}{|\delta|} \left\lfloor n' \frac{|\delta|}{|\alpha|} \right\rfloor \leq \frac{n'}{|\alpha|} \leq \frac{n(\alpha)}{|\alpha|} + \frac{2A_0}{b_j} \frac{1}{|\alpha|}$$

$$< \frac{n(\alpha)}{|\alpha|} + n(\alpha) \frac{b_j^{1/2}}{|\alpha|} = \frac{n(\alpha)}{|\alpha|} (1 + b_j^{1/2}).$$

And we have proved $\lambda(\delta) \leq (1 + b_j^{1/2})\lambda(\alpha)$.

Now assume that $|\delta| > 4|I|/2^N$. Then by the construction of Π_ξ and $I(x)$, the level j' of $\mathcal{C}_\delta f(x)$ is $j' < k$. As $I(\alpha) \setminus I(x)$ is union of intervals of Π_ξ, we can apply the basic step to obtain

$$\left| \mathcal{C}_{\xi/I(\alpha)} f(x) - \mathcal{C}_\delta f(x) \right| \leq C y b_{k-1} + 2 H_\xi^* f(x) + D y b_{k-1} \Delta_\xi(x).$$

Since $\xi \in \mathcal{S}^k$ and $x \notin E_N$, we obtain the bounds

$$H_\xi^* f(x) \leq C_2 2^m y b_{k-1}^{1/2}, \qquad \Delta_\xi(x) \leq C_1 2^m b_{k-1}^{-1/2}.$$

Hence

$$\left| \mathcal{C}_{\xi/\alpha} f(x) - \mathcal{C}_\delta f(x) \right| \leq C 2^m y b_{k-1}^{1/2}.$$

Furthermore, β controls the change of frequency, thus

$$\left| \mathcal{C}_{\xi/\alpha} f(x) - \mathcal{C}_\alpha f(x) \right| \leq B(y b_{k-1} + y b_j) \leq C y b_{k-1},$$

because $|n(\xi/\alpha) - n(\alpha)| \leq 2A_0 b_j^{-1}$, and $\||f|\|_{\beta/\alpha} < y b_{k-1}$. Combining the last two equations leads us to

$$\left| \mathcal{C}_\delta f(x) - \mathcal{C}_\alpha f(x) \right| \leq C 2^m y b_{k-1}^{1/2}.$$

$\square$

Theorem 10.2 *There exists an absolute constant C such that for every special function f, $y > 0$ and $1 < p < +\infty$, we have*

$$\mathfrak{m}\{x \in I/2 : C_I^* f(x) > y\} \leq B_p^p \frac{\|f\|_p^p}{y^p},$$

where $B_p \leq Cp^2/(p-1)$.

Proof. Given f, p and $y > 0$ we consider a natural number N (we are interested only in big values of N, $N \geq N_0$). With these elements we perform the Carleson analysis of the function f and define the shift m and the exceptional set $E_N = D \cup S^* \cup T_N \cup U_N \cup V \cup X^* \cup Y$. As we have seen in chapter eight, $|f|$ being a characteristic function

$$\mathfrak{m}(E_N) \leq C \frac{\|f\|_p^p}{y^p}.$$

We are going to prove that for every $x \in I/2 \smallsetminus E_N$

$$\sup_{0 \leq n < \theta 2^N} |C_{(n,I)} f(x)| \leq B 2^m y, \tag{10.2}$$

where θ is an absolute constant, and m is the shift considered in the previous chapters. We know that $2^m \sim Cp^2/(p-1)$.

Let $E = \{x \in I/2 : \sup_{n \geq 0} |C_{(n,I)} f(x)| > B 2^m y\}$. It is easy to see that $E \subset \liminf_N E_N$, hence we get

$$\mathfrak{m}(E) \leq C \frac{\|f\|_p^p}{y^p}.$$

If we apply the same reasoning to $\overline{f}$ we obtain an analogous inequality for the set $E' = \{x \in I/2 : \sup_{n \leq 0} |C_{(n,I)} f(x)| > B 2^m y\}$. Thus we get

$$\mathfrak{m}\{x \in I/2 : C_I^* f(x) > B 2^m y\} \leq 2C \frac{\|f\|_p^p}{y^p}.$$

From which the assertion of the theorem follows easily.

Next, we show (10.2). Take $\alpha = (n, I)$ with $0 \leq n < \theta 2^N$, $x \in I/2 \smallsetminus E_N$ and proceed to show that $|C_\alpha f(x)| \leq B 2^m y$. We start an interative process that takes a Carleson integral $C_\alpha f(x)$ of level j and obtain: a number $1 \leq k \leq j$, and a bound of $C_\alpha f(x)$, or a second Carleson integral $C_\delta f(x)$ of level $j' < k$, and besides, a good bound of the difference $|C_\delta f(x) - C_\alpha f(x)|$. Then the process continues with $C_\delta f(x)$ instead of $C_\alpha f(x)$.

We start with $\alpha = \alpha_0 = (n, I)$, but the same procedure is repeated with other pairs. So, now we assume a Carleson integral $C_\alpha f(x)$ of level j to be given, $|\alpha| > 4|I|/2^N$, and $x \in I/2 \smallsetminus E_N$.

Observe that since $x \notin V$, if $n(\alpha) = 0$ then $|C_\alpha f(x)| \leq B 2^m y$.

We apply Proposition 10.1, and obtain a natural number k, with $1 \le k \le j$ and one of these two possibilities:

1 If $|\mathcal{C}_\alpha f(x)| \le C2^m y b_{k-1}^{1/2}$, we stop the process, for we have the bound we are seeking.

2 In this second possibility the level j is a natural number and Proposition 10.1 gives us a pair δ with $I(\delta) \subsetneq I(\alpha)$, $x \in I(\delta)/2$, and $\lambda(\delta) \le (1+b_j^{1/2})\lambda(\alpha)$. We divide this case into two other cases depending on the value of $n(\alpha)$.

2.1 If $n(\alpha) = 0$, we change the value of k to $k = 1$ and we have $|\mathcal{C}_\alpha f(x)| \le C2^m y b_{k-1}^{1/2} = C2^m y$. We stop the process at α.

2.2 If $n(\alpha) > 0$ we divide it into two subcases.

2.2.1 $|\delta| \le 4|I|/2^N$. In this case we stop the process. But observe that in this case we have not attained our objective, that is, to bound $|\mathcal{C}_\alpha f(x)|$ or the difference $|\mathcal{C}_\delta f(x) - \mathcal{C}_\alpha f(x)|$.

2.2.2 We have $|\delta| > 4|I|/2^N$. Since we are now in case (B) of Proposition 10.1, the level of $\mathcal{C}_\delta f(x)$ is $j' < k$, and

$$|\mathcal{C}_\delta f(x) - \mathcal{C}_\alpha f(x)| \le C2^m y b_{k-1}^{1/2}.$$

The process starts with $\alpha = \alpha_0$, of level j_0 and such that $0 \le n(\alpha_0) < \theta 2^N$. We obtain k_0, with $1 \le k_0 \le j_0$, and stop or obtain $\alpha_1 = \delta$, such that the level of $\mathcal{C}_{\alpha_1} f(x)$ is $j_1 = j' < k_0$, and

$$|\mathcal{C}_{\alpha_1} f(x) - \mathcal{C}_{\alpha_0} f(x)| \le C2^m y b_{k_0-1}^{1/2}.$$

If the process is not stopped, $|\alpha_1| > 4|I|/2^N$ and we can continue with $\mathcal{C}_{\alpha_1} f(x)$ in the same way we have proceeded with $\mathcal{C}_{\alpha_0} f(x)$. Then we stop the process or define a new α_2, j_2 and k_2.

This process must terminate because the position of x implies that the level is always a natural number, but $j_0 > j_1 > j_2 > \cdots$

We can repeat the reasoning to obtain a sequence of pairs

$$\alpha_0, \quad \alpha_1, \quad \alpha_2, \quad \ldots \quad \alpha_s.$$

and the corresponding numbers

$$j_0 \ge k_0 > j_1 \ge k_1 > j_2 \ge k_2 > j_3 \ge \ldots > j_s \ge k_s.$$

Now when we stop the process we shall have

$$|\mathcal{C}_{\alpha_s} f(x)| \le C2^m y b_{k_s-1}^{1/2}.$$

This is because the situation depicted in **2.2.1** never happens. In fact, before that can happen we stop the process at the point **2.1**. The reason for this is connected with the selection of θ. Assume that α_s exists. We have $\lambda(\alpha_{r+1}) \le (1+b_{j_r}^{1/2})\lambda(\alpha_r)$. Therefore

$$\lambda(\alpha_s) \le \prod_{\iota=0}^{s-1}(1 + b_{j_\iota}^{1/2})\lambda(\alpha_0) \le \prod_{j=1}^{+\infty}\left(1 + b_j^{1/2}\right)\lambda(\alpha_0) = C\lambda(\alpha_0).$$

Take θ so that $\theta C < 1/4$. We have

$$2\pi\,\frac{n(\alpha_s)}{|\alpha_s|} \le 2\pi C\,\frac{n(\alpha_0)}{|\alpha_0|}.$$

Hence, if $n(\alpha_s) \ne 0$, then $n(\alpha_s) \ge 1$, and we get

$$|\alpha_s| \ge \frac{1}{C}\frac{|\alpha_0|}{n(\alpha_0)} \ge \frac{1}{C}\frac{|\alpha_0|}{\theta 2^N} > \frac{4|I|}{2^N}.$$

Therefore we arrive at $n(\alpha_s) = 0$ before we arrive at $|\alpha_s| \le 4|I|/2^N$.

The Carleson integrals $\mathcal{C}_{\alpha_\iota} f(x)$ satisfy the inequalities

$$\left|\mathcal{C}_{\alpha_{\iota+1}} f(x) - \mathcal{C}_{\alpha_\iota} f(x)\right| \le C2^m y b_{k_\iota-1}^{1/2}, \qquad 0 \le \iota \le s - 1.$$

Since we have excluded the only case where this is not true, the last integral $\mathcal{C}_{\alpha_s} f(x)$ is bounded by $\left|\mathcal{C}_{\alpha_s} f(x)\right| \le C2^m y b_{k_s-1}^{1/2}$.

Collecting these results we have

$$\left|\mathcal{C}_{\alpha_0} f(x)\right| \le \sum_{\iota=0}^{s-1}\left|\mathcal{C}_{\alpha_{\iota+1}} f(x) - \mathcal{C}_{\alpha_\iota} f(x)\right| + \left|\mathcal{C}_{\alpha_s} f(x)\right|$$

$$\le C\,2^m y \sum_{\iota=0}^{s} b_{k_\iota-1}^{1/2} \le C\,2^m y \sum_{k=1}^{\infty} b_k^{1/2} = B2^m y.$$

$\square$

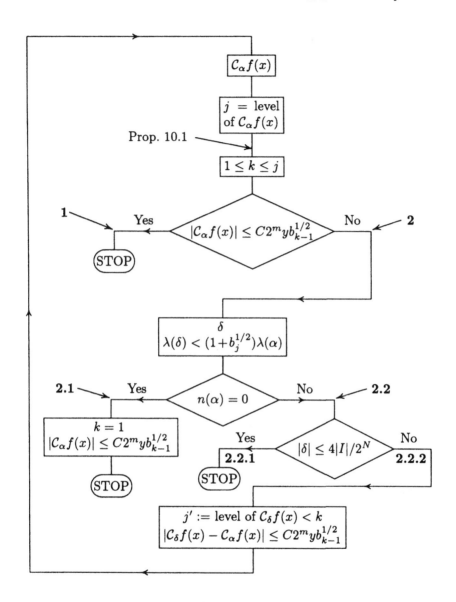

Flow diagram of the proof

Part Three

Consequences

11. Spaces of functions

11.1 Introduction

Even if we are only interested in $S^* f$, when $f \in \mathcal{L}^p$, we shall need some other spaces of functions. Since these spaces are not usually included in a first course of real analysis we give here a brief summary of their definitions and properties. The reader can find a more complete study in the references Hunt [22] and Bennett and Sharpley [3].

We will define the Lorentz spaces $\mathcal{L}^{p,1}(\mu)$ and $\mathcal{L}^{p,\infty}(\mu)$. In Theorem 11.6, we prove that $\mathcal{L}^{p,1}(\mu)$ is the smallest rearrangement invariant space of measurable functions such that $\| \chi_M \| = \| \chi_M \|_p$. We express these property by saying that it is an atomic space. Then we see $\mathcal{L}^{p,\infty}(\mu)$ as its dual space, this dictates our selection of the norm $\| \cdot \|_{p,\infty}$. The proof that we presents of the duality Theorem 11.7 may appear unnecessarilly complicated, but it has the merit of getting absolute constants.

We presents Marcinkiewicz's Theorem with special attention to the constant that appear that will have a role in our theorems on Fourier series.

We end the chapter studying a class of spaces near $\mathcal{L}^1(\mu)$ that play a prominent role in the next chapter. We prove that they are atomic spaces, a fact that allows very neat proof in the following chapter.

11.2 Decreasing rearrangement

The functions we are considering will be defined on an interval X of $\mathbf{R}$ and we will always consider the normalized Lebesgue measure on this interval, μ. Therefore μ is a probability measure $\mu(X) = 1$.

Given a measurable function $f : X \to \mathbf{R}$ we consider its **distribution function**

$$\mu_f(y) = \mu\{|f| > y\}.$$

$\mu_f : [0, +\infty) \to [0, 1]$ is a decreasing and right-continuous function.

Observe that if (f_n) is a sequence of measurable functions such that $|f_n|$ is an increasing sequence converging to $|f|$, then μ_{f_n} is increasing and converges to μ_f.

Apart from μ_f we also consider the **decreasing rearrangement** of f that is defined as

$$f^*(t) = \mathfrak{m}\{y > 0 : \mu_f(y) > t\}.$$

If f is a positive simple function, there is a decreasing finite sequence of measurable sets $(A_j)_{j=1}^n$ and positive real numbers s_j such that $f = \sum_{j=1}^n s_j \chi_{A_j}$. (If the non null values that f attains are $a_1 > a_2 > \cdots > a_n$, and $a_{n+1} = 0$ we can take $A_j = \{f > a_{j+1}\}$ and $s_j = a_j - a_{j+1}$). Then it is easy to see that $f^* = \sum_{j=1}^n s_j \chi_{[0,\mu(A_j))}$.

Proposition 11.1 *For every measurable function $f: X \to [0,+\infty]$ and every measurable set A*

$$\int_A f \, d\mu \leq \int_0^{\mu(A)} f^*(t) \, dt.$$

Proof. Since $|f| \leq |g|$ implies $f^* \leq g^*$ we only need to prove the case where f is a simple function. Then with the above representation we get

$$\int_A f \, d\mu = \sum_{j=1}^n s_j \mu(A \cap A_j); \qquad \int_0^{\mu(A)} f^*(t) \, dt = \sum_{j=1}^\infty s_j \inf\{\mu(A_j), \mu(A)\}.$$

The comparison between these quantities is trivial. $\square$

Theorem 11.2 (Hardy and Littlewood) *For all measurable functions f and $g: X \to C$ we have*

$$\left| \int_X fg \, d\mu \right| \leq \int_0^1 f^*(t) g^*(t) \, dt.$$

Proof. Since $(|f|)^* = f^*$ we can assume that f and g are positive.

First assume that f is a simple function and consider its representation as above. We shall have

$$\int fg \, d\mu = \sum_{j=1}^n s_j \int_{A_j} g \, d\mu \leq \sum_{j=1}^n s_j \int_0^{\mu(A_j)} g^*(t) \, dt = \int_0^1 f^*(t) g^*(t) \, dt.$$

For a general f let (f_n) be an increasing sequence of positive simple functions converging to f. We shall have

$$\int f_n g \, d\mu \leq \int_0^1 f_n^* g^* \, d\mathfrak{m} \leq \int_0^1 f^* g^* \, d\mathfrak{m}.$$

And now we can apply the monotone convergence theorem. $\square$

Proposition 11.3 *A measurable function* $f: X \to \mathbf{C}$ *is* **equimeasurable** *with* f^*, *that is, for every* $y > 0$ *we have* $\mu\{|f| > y\} = \mathrm{m}\{f^* > y\}$.

Proof. Since f^* is decreasing $\mathrm{m}\{f^* > y\} = \sup\{t : f^*(t) > y\}$. By the same reasoning $f^*(t) = \mathrm{m}\{\mu_f(s) > t\} = \sup\{s : \mu_f(s) > t\}$. Hence we get

$$\mathrm{m}\{f^* > y\} = \sup\{t : \text{ there is } s > y, \text{ with } \mu_f(s) > t\}.$$

Since $\mu\{f > s_n\} \to \mu\{f > y\}$ for every decreasing sequence (s_n) converging to y, we get

$$\mathrm{m}\{f^* > y\} = \sup\{t : \mu_f(y) > t\} = \mu_f(y).$$

$\square$

Proposition 11.4 *If* φ *is a nonnegative, Borel measurable function on* $[0, +\infty)$, *we have*

$$\int_X \varphi(|f|)\, d\mu = \int_0^1 \varphi(f^*(t))\, dt,$$

for every measurable function f.

Proof. Let ν_1 be the Borel measure on $[0, +\infty)$ image of μ by the function $|f|$, and let ν_1 the Borel measure on $[0, +\infty)$ image of m by the function f^*.

The two functions $|f|$ and f^* are equimeasurable

$$\mathrm{m}\{t \in (0, 1) : f^*(t) > s\} = \mu\{x \in X : |f(x)| > s\}.$$

Therefore, for $s > 0$, we have $\nu_1(s, +\infty) = \nu_2(s, +\infty)$. Since ν_1 and ν_2 are probabilities we have also $\nu_1\{0\} = \nu_2\{0\}$. Therefore the two image measures are the same.

Then since φ is Borel measurable and positive we have

$$\int_X \varphi(|f|)\, d\mu = \int_{[0, +\infty)} \varphi\, d\nu_1 = \int_{[0, +\infty)} \varphi\, d\nu_2 = \int_0^1 \varphi(f^*)\, d\mathrm{m}.$$

$\square$

For every measurable function $f: X \to \mathbf{C}$ we define for $t > 0$ $f^{**}(t)$ in the following way

$$f^{**}(t) = \frac{1}{t} \int_0^t f^*(s)\, ds.$$

This function is also the supremum of the mean values of f.

Proposition 11.5 *For every measurable function*

$$f^{**}(t) = \sup\left\{\frac{1}{\mu(A)} \int_A |f|\, d\mu : \mu(A) = t\right\}.$$

Proof. By Proposition 11.1, for every measurable set A with $\mu(A) = t$

$$\int_A |f| \, d\mu \leq \int_0^t f^*(s) \, ds$$

Therefore the supremum of the mean values of $|f|$ is less than or equal to $f^{**}(t)$.

To prove the equality, first assume that there is an $y > 0$ such that $\mu\{|f| > y\} = m\{f^* > y\} = t$. Then we apply Theorem 11.4 with $\varphi(t) = t\chi_{(y,+\infty)}(t)$. If we denote by A the set $\{|f| > y\}$ we get

$$\int_A |f| \, d\mu = \int \varphi(|f|) \, d\mu = \int \varphi(f^*) \, dm = \int_0^t f^* \, dm.$$

In the other case there is some point y_0 such that $\mu\{|f| > y\} < t$ for $y > y_0$ and $\mu\{|f| > y\} > t$ for $y < y_0$. Then there is a set of positive measure on X where $|f| = y_0$. Hence f^* takes this value on a set of positive measure. Then we can obtain a set $A = A_0 \cup A_1$ where $A_0 = \{|f| > y_0\}$ and $A_1 \subset \{|f| = y_0\}$ and such that $\mu(A) = t$. It is easy to see that in this case we also obtain the equality.   □

Therefore we have $(f + g)^{**}(t) \leq f^{**}(t) + g^{**}(t)$.

11.3 The Lorentz spaces $\mathcal{L}^{p,1}(\mu)$ and $\mathcal{L}^{p,\infty}(\mu)$

First we consider the Lorentz spaces $\mathcal{L}^{p,1}(\mu)$. It is the set of those measurable functions $f: X \to \mathbf{C}$ such that

$$\|f\|_{p,1} = \frac{1}{p} \int_0^1 t^{1/p} f^*(t) \frac{dt}{t} = \int_0^1 f^*(t^p) \, dt < +\infty.$$

Every function $f \in \mathcal{L}^{p,1}(\mu)$ is in $\mathcal{L}^p(\mu)$. In the case that $p = 1$ there is nothing to prove. For $p > 1$ let q be the conjugate exponent. Then

$$\|f\|_p = \sup_{\|g\|_q \leq 1} \left| \int fg \, d\mu \right| \leq \int_0^1 f^* g^* \, dm.$$

It is easy to see that for every such g, we have $g^*(t) \leq t^{-1/q}$. Therefore we get

$$\|f\|_p \leq p\|f\|_{p,1}.$$

It can be proved that the above inequality can be improved, removing the coefficient p.

Now we show that $\|f\|_{p,1}$ is a norm and $\mathcal{L}^{p,1}(\mu)$ is a Banach space. From the equality $(\lambda f)^* = |\lambda| f^*$, we get $\|\lambda f\|_{p,1} = |\lambda| \, \|f\|_{p,1}$. To prove the triangle inequality we bring in f^{**},

$$\|f+g\|_{p,1} = p^{-1} \int_0^1 t^{-1/q}(f+g)^* \, dt$$

$$= p^{-1} t^{-1/q} t(f+g)^{**} \Big|_0^1 + p^{-1} q^{-1} \int_0^1 t(f+g)^{**} t^{-1-1/q} \, dt.$$

Since $(f+g) \in \mathcal{L}^p$ we know that $\lim_{t\to 0^+} t^{1/p}(f+g)^{**}(t) = 0$. Hence

$$\|f+g\|_{p,1} = p^{-1}(f+g)^{**}(1) + p^{-1} q^{-1} \int_0^1 (f+g)^{**} t^{-1/q} \, dt.$$

We have seen that $(f+g)^{**} \leq f^{**} + g^{**}$. It follows that $\|\cdot\|_{p,1}$ is a norm.

To prove that the space is complete it suffices to prove that for every sequence of functions (f_j) with $\sum_{j=1}^\infty \|f_j\|_{p,1} < +\infty$, the series $\sum_{j=1}^\infty f_j$ is absolutely convergent almost everywhere on X, and that the sum S of the series satisfies $\|S\|_{p,1} \leq \sum_{j=1}^\infty \|f_j\|_{p,1}$.

To prove this assertion we can assume that the functions $f_j \geq 0$ for every $j \in \mathbf{N}$. Then the partial sums S_j are increasing, therefore S_j^* is increasing and converges to S^*. Hence

$$p^{-1} \int_0^1 t^{-1/q} S^*(t) \, dt = \lim_{N\to +\infty} p^{-1} \int_0^1 t^{-1/q} S_N^*(t) \, dt \leq \sum_{j=1}^\infty \|f_j\|_{p,1}.$$

For a characteristic function χ_A a simple computation shows that $\|\chi_A\|_{p,1}$ is equal to $\|\chi_A\|_p$. The space $\mathcal{L}^{p,1}(\mu)$ $(p > 1)$ can be defined as the smallest Banach space with this property. This is the content of the following theorem.

Theorem 11.6 *There is an absolute constant C such that for every $f \in \mathcal{L}^{p,1}(\mu)$ there exist a sequence of measurable sets $(A_j)_{j=1}^\infty$ and numbers $(a_j)_{j=1}^\infty$ such that*

$$f = \sum_{j=1}^\infty a_j \chi_{A_j}, \qquad \|f\|_{p,1} \leq \sum_{j=1}^\infty |a_j| \mu(A_j)^{1/p} \leq C\|f\|_{p,1}.$$

Proof. Given the measurable function f, by induction we can define a partition of X into measurable sets $(A_j)_{j=1}^\infty$ such that $\mu(A_j) = (e^p - 1)e^{-pj}$ and for every $x \in A_j$, $y \in A_k$ with $j < k$ we have $|f(x)| \leq |f(y)|$. Define $a_j = \sup_{x\in A_j} |f(x)|$.

We can easily check that $f^* \geq \sum_{j=1}^\infty a_j \chi_{I_j}$, were I_j denotes the interval $I_j = [e^{-p(j+1)}, e^{-pj})$. Therefore

$$\|f\|_{p,1} \geq (1 - e^{-1}) \sum_{j=1}^\infty a_j e^{-j}.$$

By the construction of A_j we have $a_1 \leq a_2 \leq \cdots$. If we assume that f is not equivalent to 0, there exists some N such that $a_1 = \cdots = a_{N-1} = 0$ and $a_N > 0$. Then we can write

$$f = \sum_{j=N}^{\infty} a_j f_j, \quad \text{where} \quad f_j = a_j^{-1} f \chi_{A_j}, \quad \|f_j\|_\infty \leq 1.$$

Since every function with values in $[0,1)$ can be writen as $\sum_{j=1}^{\infty} 2^{-j} \chi_{T_j}$, we get

$$f_j = \sum_{k=1}^{\infty} \beta_{j,k} \chi_{T_{j,k}}, \quad T_{j,k} \subset A_j, \quad \sum_{k=1}^{\infty} |\beta_{j,k}| \leq 4.$$

Therefore we obtain the expression

$$f = \sum_{j,k=1}^{\infty} a_j \beta_{j,k} \chi_{T_{j,k}}.$$

This is the decomposition we are seeking. In fact

$$\sum_{j,k} a_j |\beta_{j,k}| \mu(T_{j,k})^{1/p} \leq 4 \sum_{j=1}^{\infty} a_j \mu(A_j)^{1/p} = 4(e^p - 1)^{1/p} \sum_{j=1}^{\infty} a_j e^{-j}$$

$$\leq \frac{4e}{1 - e^{-1}} \|f\|_{p,1}.$$

$\square$

In particular we have proved the density of the simple functions in the space $\mathcal{L}^{p,1}(\mu)$.

Remark. I stress the fact that C is an absolute constant in the previous theorem. It follows, for example that the inequality $\|f\|_p \leq p\|f\|_{p,1}$ obtained above, can be improved now. In fact, under the conditions of the theorem, we have $\|f\|_p \leq \sum |a_j| \|\chi_{A_j}\|_p \leq C\|f\|_{p,1}$.

Now we consider the dual space. We define $\mathcal{L}^{p,\infty}(\mu)$ as the set of measurable functions f such that $\sup_{0<t<1} t^{1/p} f^*(t) < +\infty$.

Since f^* is decreasing, by definition, $f^{**}(t) \geq f^*(t)$. Therefore

$$\sup_{0<t<1} t^{1/p} f^*(t) \leq \sup_{0<t<1} t^{1/p} f^{**}(t) := \|f\|_{p,\infty}.$$

For $p > 1$, these two quantities are equivalent. Indeed if we assume

$$\sup_{0<t<1} t^{1/p} f^*(t) = C,$$

then

$$f^{**}(t) = \frac{1}{t}\int_0^t f^*(s)\,ds \le \frac{Cp}{p-1}t^{-1/p} \qquad (11.1).$$

It is easy to see that as we have defined it $\|\cdot\|_{p,\infty}$ is a norm, and $\mathcal{L}^{p,\infty}(\mu)$ a vector space. For $p=1$, we can see that $\mathcal{L}^{1,\infty}(\mu)$ is a vector space. It is the weak $\mathcal{L}_1$ space. It can be shown that it is not a normed space. Therefore for $p=1$ we put

$$\|f\|_{1,\infty} = \sup_{0<t<1} t^{1/p}f^*(t).$$

This is not a norm but a quasi-norm.

Now it is straightforward to see that for $p>1$ the space $\mathcal{L}^{p,\infty}(\mu)$ is a Banach space. Also it is clear that $\mathcal{L}^p(\mu) \subset \mathcal{L}^{p,\infty}(\mu)$. In fact, by Hölder's inequality,

$$t^{1/p}f^{**}(t) = t^{1/p-1}\int_0^t f^*(s)\,ds \le t^{1/p-1}t^{1/q}\|f^*\|_p = \|f\|_p.$$

Therefore

$$\|f\|_{p,\infty} \le \|f\|_p.$$

Theorem 11.7 *For $p>1$, $\mathcal{L}^{q,\infty}(\mu)$ is the dual space of $\mathcal{L}^{p,1}(\mu)$, where q is the conjugate exponent to p.*

Proof. Given $g \in \mathcal{L}^{q,\infty}(\mu)$, for every $f \in \mathcal{L}^{p,1}(\mu)$ we put $f = \sum_{j=1}^\infty a_j\chi_{A_j}$, with $\sum_{j=1}^\infty |a_j|\|\chi_{A_j}\|_p \le C\|f\|_{p,1}$. Then

$$\left|\int_X fg\,d\mu\right| \le \sum_{j=1}^\infty |a_j|\int_{A_j}|g|\,d\mu \le \sum_{j=1}^\infty |a_j|\mu(A_j)g^{**}(\mu(A_j))$$

$$\le \sum_{j=1}^\infty |a_j|\mu(A_j)^{1-1/q}\|g\|_{q,\infty} \le \sum_{j=1}^\infty |a_j|\|\chi_{A_j}\|_p\|g\|_{q,\infty}$$

$$\le C\|f\|_{p,1}\|g\|_{q,\infty}.$$

This allows us to identify every function $g \in \mathcal{L}^{q,\infty}(\mu)$ with a continuous linear functional defined in $\mathcal{L}^{p,1}(\mu)$.

Let u be a continuos linear functional on $\mathcal{L}^{p,1}(\mu)$. Put $\nu(A) = u(\chi_A)$, for every measurable set A. Then ν is an additive set function. Since $|\nu(A)| \le \|u\|\|\chi_A\|_{p,1} = \|u\|\mu(A)^{1/p}$, the function ν is a signed measure and $\nu \ll \mu$. By the Radon-Nikodym Theorem there is a measurable function g, such that $\nu(A) = \int_A g\,d\mu$, for every measurable set A. This function is in $\mathcal{L}^{q,\infty}(\mu)$. In fact we have

$$\left|\int_A g\,d\mu\right| \le \|u\|\mu(A)^{1/p}.$$

From this we derive that $g^{**}(t) \le \|u\|t^{-1/q}$. Therefore $\|g\|_{q,\infty} \le \|u\|$.

It follows that for every simple function φ we have $u(\varphi) = \int \varphi g \, d\mu$. By the density of the simple functions and the inequality $\left| \int fg \, d\mu \right| \leq C \|f\|_{p,1} \|g\|_{q,\infty}$, it follows that $u(f) = \int fg \, d\mu$ for every $f \in \mathcal{L}^{p,1}(\mu)$.

Therefore $\mathcal{L}^{q,\infty}(\mu)$ is the dual space. Also, if we denote by $\|\cdot\|_{p,\infty}^{\bullet}$ the dual norm in the space $\mathcal{L}^{q,\infty}(\mu)$,

$$\|g\|_{q,\infty}^{\bullet} = \sup_{\|f\|_{p,1} \leq 1} \left| \int_X fg \, d\mu \right|.$$

Then we have proved that for some absolute constant C we have

$$\|g\|_{q,\infty} \leq \|g\|_{q,\infty}^{\bullet} \leq C \|g\|_{q,\infty}.$$

$\square$

11.4 Marcinkiewicz Interpolation Theorem

To prove this interpolation theorem we shall need the following inequality.

Theorem 11.8 (Hardy's inequality) *For every positive real function f defined on $(0, +\infty)$, and every $1 \leq p_1 < p < p_0 < +\infty$ we have*

$$\left(\int_0^\infty t^{-\frac{p}{p_1}} \left(\int_0^t f(s) s^{\frac{1}{p_1}-1} \, ds \right)^p dt \right)^{\frac{1}{p}} \leq \left(\frac{1}{p_1} - \frac{1}{p} \right)^{-1} \left(\int_0^\infty f(s)^p \, ds \right)^{\frac{1}{p}},$$

$$\left(\int_0^\infty t^{-\frac{p}{p_0}} \left(\int_t^{+\infty} f(s) s^{\frac{1}{p_0}-1} \, ds \right)^p dt \right)^{\frac{1}{p}} \leq \left(\frac{1}{p} - \frac{1}{p_0} \right)^{-1} \left(\int_0^\infty f(s)^p \, ds \right)^{\frac{1}{p}}.$$

Proof. Both inequalities are proved in the same way. For example, to prove the first one, we apply Hölder's inequality to the inner integral

$$\int_0^t f(s) s^{-\frac{1}{p'}} s^{\frac{1}{p_1}-\frac{1}{p}} \, ds \leq \left(\int_0^t f(s)^p s^{\frac{1}{p_1}-\frac{1}{p}} \, ds \right)^{\frac{1}{p}} \left(\frac{t^{\frac{1}{p_1}-\frac{1}{p}}}{\frac{1}{p_1}-\frac{1}{p}} \right)^{1-\frac{1}{p}}.$$

Then, if we denote by I the first term of the inequality, we get

$$I^p \leq \left(\frac{1}{p_1} - \frac{1}{p} \right)^{1-p} \int_0^\infty \left(\int_0^t f(s)^p s^{1/p_1-1/p} \, ds \right) t^{1/p-1/p_1-1} \, dt$$

After applying Fubini's Theorem we get

$$I^p \leq \left(\frac{1}{p_1} - \frac{1}{p} \right)^{1-p} \int_0^{+\infty} f(s)^p s^{1/p_1-1/p} \left(\int_s^{+\infty} t^{1/p-1/p_1-1} \, dt \right) ds$$

$$\leq \left(\frac{1}{p_1} - \frac{1}{p} \right)^{-p} \int_0^\infty f(s)^p \, ds.$$

□

Now we are ready to prove Marcinkiewicz's Theorem. We say that an operator S mapping a vector space of measurable functions to measurable functions is **sublinear** if $|S(f + g)| \leq |Sf| + |Sg|$, and for every scalar a, we have $|S(af)| = |a||Sf|$.

Theorem 11.9 (Marcinkiewicz) *Let S be a sublinear operator defined on $\mathcal{L}^{p_0,1}(\mu) + \mathcal{L}^{p_1,1}(\mu)$ where $+\infty > p_0 > p_1 > 1$. Assume that there exist constants M_0 and M_1 such that*

$$\|Sf\|_{p_0,\infty} \leq M_0\|f\|_{p_0,1}, \quad and \quad \|Sf\|_{p_1,\infty} \leq M_1\|f\|_{p_1,1}.$$

Then, for every $p \in (p_1, p_0)$, $S: \mathcal{L}^p(\mu) \to \mathcal{L}^p(\mu)$ is continuous with norm

$$\|S\|_p \leq \frac{p(p_0 - p_1)}{(p_0 - p)(p - p_1)} M_0^{1-\theta} M_1^\theta, \quad where \quad \frac{1}{p} = \frac{1 - \theta}{p_0} + \frac{\theta}{p_1}.$$

Proof. Let f be a function in $\mathcal{L}^p(\mu)$. First we bound $(Sf)^{**}(t)$ using the hypotheses that S maps $\mathcal{L}^{q,1}(\mu) \to \mathcal{L}^{q,\infty}(\mu)$ for $q = p_0$ and $q = p_1$. We decompose f into two functions, $f = f_0 + f_1$, in the following way

$$f_0(s) = \begin{cases} f(s) & \text{if } |f(s)| \leq f^*(at), \\ f^*(at)\,\mathrm{sgn}(f(s)) & \text{if } |f(s)| \geq f^*(at); \end{cases}$$

$$f_1(s) = \begin{cases} 0 & \text{if } |f(s)| \leq f^*(at), \\ f(s) - f^*(at)\,\mathrm{sgn}(f(s)) & \text{if } |f(s)| \geq f^*(at). \end{cases}$$

Here a is a parameter that we will choose later. It is easy to see that the decreasing rearrangements of these functions are given by

$$f_0^*(s) = \begin{cases} f^*(s) & \text{if } s \geq at \\ f^*(at) & \text{if } s \leq at \end{cases} \qquad f_1^*(s) = \begin{cases} 0 & \text{if } s \geq at \\ f^*(s) - f^*(at) & \text{if } s \leq at \end{cases}$$

Since $f_0 \in \mathcal{L}_{p_0,1}$ and $f_1 \in \mathcal{L}_{p_1,1}$ we see that S is defined on $\mathcal{L}^p$.

Since we have

$$(Sf)^{**}(t) \leq (Sf_0)^{**}(t) + (Sf_1)^{**}(t) \leq$$

$$\frac{M_0}{p_0} t^{-1/p_0} \int_0^1 f_0^*(s) s^{1/p_0 - 1}\, ds + \frac{M_1}{p_1} t^{-1/p_1} \int_0^1 f_1^*(s) s^{1/p_1 - 1}\, ds,$$

$(Sf)^{**}(t)$ is bounded by

$$(M_0 a^{1/p_0} - M_1 a^{1/p_1}) f^*(at) + \frac{M_0}{p_0} t^{-1/p_0} \int_{at}^1 f^*(s) s^{1/p_0 - 1}\, ds$$

$$+ \frac{M_1}{p_1} t^{-1/p_1} \int_0^{at} f^*(s) s^{1/p_1 - 1}\, ds.$$

We consider $f^*(s)$, defined for $s > 1$, equal to 0.

By Proposition 11.4, $\|Sf\|_p = \|Sf^*\|_p$, and since $Sf^* \leq Sf^{**}$, we obtain

$$\|Sf\|_p \leq \|(Sf)^{**}\|_p \leq (M_0 a^{1/p_0} - M_1 a^{1/p_1}) a^{-1/p} \|f\|_p$$
$$+ \frac{M_0}{p_0} \left(\int_0^\infty t^{-p/p_0} \left(\int_{at}^1 f^*(s) s^{1/p_0 - 1}\, ds \right)^p dt \right)^{1/p} +$$
$$\frac{M_1}{p_1} \left(\int_0^\infty t^{-p/p_1} \left(\int_0^{at} f^*(s) s^{1/p_1 - 1}\, ds \right)^p dt \right)^{1/p}.$$

With a change of variables we get

$$\|Sf\|_p \leq (M_0 a^{1/p_0} - M_1 a^{1/p_1}) a^{-1/p} \|f\|_p$$
$$+ \frac{M_0}{p_0} a^{1/p_0 - 1/p} \left(\int_0^\infty t^{-p/p_0} \left(\int_t^{+\infty} f^*(s) s^{1/p_0 - 1}\, ds \right)^p dt \right)^{1/p}$$
$$+ \frac{M_1}{p_1} a^{1/p_1 - 1/p} \left(\int_0^\infty t^{-p/p_1} \left(\int_0^t f^*(s) s^{1/p_1 - 1}\, ds \right)^p dt \right)^{1/p}.$$

Now we apply Hardy's inequality. It follows that

$$\|Sf\|_p \leq (M_0 a^{1/p_0} - M_1 a^{1/p_1}) a^{-1/p} \|f\|_p$$
$$+ \frac{p M_0}{p_0 - p} a^{1/p_0 - 1/p} \left(\int_0^\infty f^*(s)^p\, ds \right)^{1/p}$$
$$+ \frac{p M_1}{p - p_1} a^{1/p_1 - 1/p} \left(\int_0^\infty f^*(s)^p\, ds \right)^{1/p}.$$

Therefore

$$\|Sf\|_p \leq \left(\frac{p_0 M_0}{p_0 - p} a^{1/p_0 - 1/p} + \frac{p_1 M_1}{p - p_1} a^{1/p_1 - 1/p} \right) \|f\|_p.$$

Now we select the best value for a. In this case $M_0 = M_1 a^{1/p_0 - 1/p_1}$ is the best choice. With this election we obtain

$$\|Sf\|_p \leq \frac{p(p_0 - p_1)}{(p_0 - p)(p - p_1)} M_0^{1-\theta} M_1^\theta, \quad \text{where} \quad \frac{1}{p} = \frac{1-\theta}{p_0} + \frac{\theta}{p_1}.$$

$\square$

Remark. In fact we have proved that

$$\|Sf\|_{p,p} = \left(\int_0^\infty f^{**}(t)^p\, dt \right)^{1/p}.$$

is bounded. This is an equivalent norm, but Hardy's inequality gives

$$\|f\|_p \leq \|f\|_{p,p} \leq \frac{p}{p-1} \|f\|_p.$$

In the above theorem we assume that $\|Sf\|_{p,\infty} \leq M\|f\|_{p,1}$. The norm $\|Sf\|_{p,\infty}$ refers to $(Sf)^{**}$ and it is usually easier to bound $(Sf)^*$. Another problem with the theorem as we have given it, is that we have excluded the case $p_1 = 1$, because in this case $\|Sf\|_{1,\infty}$ is not a norm, and is defined in another way. Therefore we give another version.

Theorem 11.10 (Marcinkiewicz) *Let S be a sublinear operator defined on $\mathcal{L}^{p_0,1}(\mu) \cup \mathcal{L}^{p_1,1}(\mu)$ where $+\infty > p_0 > p_1 \geq 1$. Assume that there exist constants M_0 and M_1 such that*

$$\sup_{0<t<1} t^{1/p_0}(Sf)^*(t) \leq M_0\|f\|_{p_0,1}, \quad and \quad \sup_{0<t<1} t^{1/p_1}(Sf)^*(t) \leq M_1\|f\|_{p_1,1}.$$

Then, for every $p \in (p_1, p_0)$, $S: \mathcal{L}^p(\mu) \to \mathcal{L}^p(\mu)$ is continuous with norm

$$\|S\|_p \leq 2^{1/p} \frac{p(p_0 - p_1)}{(p_0 - p)(p - p_1)} M_0^{1-\theta} M_1^{\theta}, \quad where \quad \frac{1}{p} = \frac{1-\theta}{p_0} + \frac{\theta}{p_1}.$$

Proof. We follow the same procedure as in the previous theorem. Instead of a bound for $(Sf)^{**}(t)$ we obtain a bound for $(Sf)^*(2t)$. It is easy to see that

$$(Sf)^*(2t) \leq (Sf_0)^*(t) + (Sf_1)^*(t).$$

Observe that $\|Sf\|_p = \|(Sf)^*\|_p$, and

$$2^{-1/p}\|(Sf)^*\|_p = \left(\int_0^\infty (Sf)^*(2t)^p \, dt\right)^{1/p} \leq \|(Sf_0)^* + (Sf_1)^*\|_p.$$

Then we follow the same reasoning as before.   □

11.5 Spaces near $\mathcal{L}^1(\mu)$

We can define many spaces between $\mathcal{L}^1(\mu)$ and $\bigcup_{p>1} \mathcal{L}^p(\mu)$. They will play a role in the problem of the almost everywhere convergence of Fourier series, since every f in the last space has an a. e. convergent Fourier series, and by Kolmogorov's example there exists a function in $\mathcal{L}^1$ whose Fourier series is everywhere divergent.

We shall define a space that we call $\mathcal{L}\varphi(\mathcal{L})$, where $\varphi: [0, +\infty) \to [0, +\infty)$ will be a function such that:
(1) There exists a constant $C > 0$ such that for every $t > 0$,

$$\varphi(t^2) \leq C\varphi(t).$$

(2) $\varphi(t)$ is absolutely continuous and $\varphi'(t) \geq 0$ a. e.
(3) $\varphi(0) = 0$.
(4) $\lim_{t \to +\infty} \varphi(t) = +\infty$.

The space $\mathcal{L}\varphi(\mathcal{L})$ will be the set of measurable functions such that

$$\int_0^{+\infty} |f|\varphi(|f|)\, d\mu < +\infty.$$

Proposition 11.11 *Assuming that φ satifies the above conditions, $\mathcal{L}\varphi(\mathcal{L})$ is a Banach space whose norm is given by*

$$\|f\|_{\mathcal{L}\varphi(\mathcal{L})} = \int_0^1 f^*(t)\varphi(1/t)\, dt = \int_0^{+\infty} \frac{f^{**}(t)}{t}\varphi'\left(\frac{1}{t}\right) dt.$$

Proof. The equality of the two expressions given for the norm is a consequence of Fubini's Theorem applied to the function

$$\frac{1}{t^2}\varphi'\left(\frac{1}{t}\right) f^*(s)\, \chi_{\{0<s<t\}},$$

in the set $[0, +\infty)^2$. (Observe that $f^*(s)$ is a positive measurable function that vanishes for $s > 1$).

Then since $\varphi' \geq 0$, the second expression and the known properties of f^{**} prove that $\|f\|_{\mathcal{L}\varphi(\mathcal{L})}$ is a norm.

Now we can prove that the norm is finite precisely in the set $\mathcal{L}\varphi(\mathcal{L})$. In fact, for a given measurable f, we define $A = \{f^*(t)^2 > 1/t\}$ and we obtain, by property (1),

$$\int_0^1 f^*(t)\varphi(1/t)\, dt = C\int_A f^*(t)\varphi(f^*(t))\, dt + \int_0^1 \frac{1}{\sqrt{t}}\varphi(1/t)\, dt$$

$$C\int |f|\varphi(|f|)\, d\mu + 2\int_0^1 \varphi(x^{-2})\, dx.$$

The last integral is finite. In fact, it is comparable to

$$\sum_{j=1}^{\infty} \frac{1}{2^n}\varphi(2^{2n}) \leq \sum \frac{e^{\alpha \log n}}{2^n} < +\infty.$$

(We have used repeatedly the condition (1)).

Therefore the norm is finite for every function in $\mathcal{L}\varphi(\mathcal{L})$.

On the other hand, if $\|f\|_{\mathcal{L}\varphi(\mathcal{L})}$, then $\|f\|_1 < +\infty$. Hence $tf^*(t) \leq \|f\|_1$. Therefore

$$\int_X |f|\varphi(|f|)\, d\mu = \int_0^1 f^*(t)\varphi(f^*(t))\, dt \leq \int_0^1 f^*(t)\varphi(\|f\|_1/t)\, dt.$$

Now if $\|f\|_1 \leq 1$ we will have

$$\int_X |f|\varphi(|f|)\, d\mu \leq \int_0^1 f^*(t)\varphi(1/t)\, dt = \|f\|_{\mathcal{L}\varphi(\mathcal{L})}.$$

And if $\|f\|_1 > 1$ we will have

$$\int_X |f|\varphi(|f|)\,d\mu \leq \|f\|_1 \int_0^1 f^*(\|f\|_1 t)\varphi(1/t)\,dt$$

$$\leq \|f\|_1 \int_0^1 f^*(t)\varphi(1/t)\,dt \leq \|f\|_1 \|f\|_{\mathcal{L}\varphi(\mathcal{L})} < +\infty.$$

The proof that $\mathcal{L}\varphi(\mathcal{L})$ is a Banach space can be given as in the $\mathcal{L}^p(\mu)$ case. Given a sequence of functions (f_n) such that $\sum_n \|f_n\|_{\mathcal{L}\varphi(\mathcal{L})} < +\infty$, we prove that the series $\sum_n f_n$ is absolutely convergent a. .e. and defines a measurable function F. Then it is easy to prove that $F = \sum_n f_n$ in the space $\mathcal{L}\varphi(\mathcal{L})$. $\square$

An important information about the space $\mathcal{L}\varphi(\mathcal{L})$ is the value of the norm of a characteristic function. We have

$$\frac{1}{2}\mu(M)\varphi(2/\mu(M)) \leq \|\chi_M\|_{\mathcal{L}\varphi(\mathcal{L})} \leq C_\varphi\,\mu(M)\varphi(2/\mu(M)).$$

This follows from the following

Lemma 11.12 *There exists a constant C_φ, such that*

$$\frac{x}{2}\varphi(2/x) \leq \int_0^x \varphi(1/t)\,dt \leq C_\varphi\, x\, \varphi(2/x), \qquad 0 < x < 1.$$

Proof. Since $\varphi(t^2) \leq C\varphi(t)$, we have

$$\frac{x}{2}\varphi(2/x) \leq \int_0^x \varphi(1/t)\,dt \leq \int_{x^2}^x \varphi(1/t)\,dt + \int_0^{x^2} \varphi(1/t)\,dt \leq$$

$$C(x - x^2)\varphi(1/x) + \int_0^{x^2} \varphi(1/t)\,dt.$$

On the other hand we have

$$\int_0^{x^2} \varphi(1/t)\,dt = 2\int_0^x \varphi(1/u^2)u\,du \leq 2C\int_0^x \varphi(1/u)u\,du$$

$$\leq 2Cx \int_0^x \varphi(1/u)\,du.$$

It follows that there exists some x_0 such that

$$\int_0^{x^2} \varphi(1/t)\,dt \leq \frac{1}{2}\int_0^x \varphi(1/t)\,dt, \qquad 0 < x < x_0.$$

Therefore in $[0, x_0]$ we have

$$\frac{x}{2}\varphi(2/x) \le \int_0^x \varphi(1/t)\,dt \le 2C\,x\,\varphi(1/x).$$

In the interval $[x_0, 1]$ the functions $x\varphi(2/x)$ and $\int_0^x \varphi(1/t)\,dt$ are continuous and non null. Therefore our lemma is true also in this interval. Observe that the conditions $\varphi(t^2) \le C\varphi(t)$ and $\lim_{t\to+\infty} \varphi(t) = +\infty$ implies that $\varphi(2) = 0$ is impossible. $\square$

A property that is important for us is that these spaces are atomic.

Theorem 11.13 *There exists a constant C_φ such that for every function $f \in \mathcal{L}\varphi(\mathcal{L})$ there exist a sequence of measurable sets $(A_j)_{j=1}^\infty$ and a sequence of complex numbers $(a_j)_{j=1}^\infty$ such that $f = \sum_{j=1}^\infty a_j \chi_{A_j}$, and*

$$\|f\|_{\mathcal{L}\varphi(\mathcal{L})} \le \sum_{j=1}^\infty |a_j|\,\|\chi_{A_j}\|_{\mathcal{L}\varphi(\mathcal{L})} \le C_\varphi \|f\|_{\mathcal{L}\varphi(\mathcal{L})}.$$

Proof. Given the measurable function f, as in Theorem 11.6, we can put $f = \sum_{j=N}^\infty a_j f_j$ where f_j has support on a measurable set A_j, $a_j \ge 0$, $\|f_j\|_\infty \le 1$ and $f^* \ge \sum_{j=1}^\infty a_j \chi_{I_j}$, with $I_j = [e^{-j-1}, e^{-j})$ and $\mu(A_j) = (e-1)e^{-j} = m_j$.
As in Theorem 11.6 we obtain the decomposition

$$f = \sum_{j,k} a_j \beta_{j,k} \chi_{T_{j,k}}.$$

Then we have

$$\sum_{j,k} |a_j \beta_{jk}| \|\chi_{T_{j,k}}\|_{\mathcal{L}\varphi(\mathcal{L})} \le 4 \sum_{j=1}^\infty a_j \int_0^{\mu(A_j)} \varphi(1/t)\,dt$$

$$\le 4C \sum_{j=1}^\infty a_j m_j \varphi(2/m_j) \le 4C(e-1) \sum_{j=1}^\infty a_j e^{-j} \varphi\big(2(e-1)^{-1}e^j\big)$$

$$\le C' \sum_{j=1}^\infty a_j e^{-j}(1-e^{-1})\varphi(e^{j+1}) \le C' \sum_{j=1}^\infty a_j \int_{I_j} \varphi(1/t)\,dt$$

$$\le C' \int_0^1 f^*(t)\varphi(1/t)\,dt = C'\|f\|_{\mathcal{L}\varphi(\mathcal{L})}.$$

$\square$

11.6 The spaces $\mathcal{L} \log \mathcal{L}(\mu)$ and $\mathcal{L}_{\exp}(\mu)$

The space $\mathcal{L} \log \mathcal{L}(\mu)$ is the space of measurable functions f such that $\int_X |f| \log^+ |f| \, d\mu < +\infty$.

By the previous section we know that

$$\|f\|_{\mathcal{L} \log \mathcal{L}} = \int_0^1 f^{**}(t) \, dt,$$

is a norm in this space.

That the above expression is a norm in the vector space where it is finite is a trivial consequence of the inequality $(f + g)^{**} \leq f^{**} + g^{**}$. By Fubini's Theorem we obtain that

$$\|f\|_{\mathcal{L} \log \mathcal{L}} = \int_0^1 f^*(t) \log(1/t) \, dt.$$

Since $f^* \leq f^{**}$ we have $\|f\|_1 \leq \|f\|_{\mathcal{L} \log \mathcal{L}}$.

We shall need the expression of the norm of a characteristic function

$$\| \chi_A \|_{\mathcal{L} \log \mathcal{L}} = \mu(A) \log\big(e/\mu(A)\big).$$

By Theorem 11.13 we get the following.

Theorem 11.14 *There is an absolute constant C such that for every function $f \in \mathcal{L} \log \mathcal{L}$ there exists a sequence of measurable sets $(A_j)_{j=1}^\infty$ and a sequence of complex numbers (a_j) such that*

$$f = \sum_{j=1}^\infty a_j \chi_{A_j}, \quad \text{and} \quad \|f\|_{\mathcal{L} \log \mathcal{L}} \leq \sum_{j=1}^\infty |a_j| \| \chi_A \|_{\mathcal{L} \log \mathcal{L}} \leq C \|f\|_{\mathcal{L} \log \mathcal{L}}.$$

The dual space of $\mathcal{L} \log \mathcal{L}$ can be characterized in several ways. We shall call it $\mathcal{L}_{\exp}$. It is the space of measurable functions such that $\mu\{|f| > y\} \leq A e^{-By}$ for some constants A and B (that change with f). It is also the space of functions such that there is some positive real number α such that $\int e^{\alpha |f|} \, d\mu < +\infty$. The equivalence of these two conditions is easy to prove. But to obtain a norm that defines the space we define the space as the set of measurable functions such that

$$\|f\|_{\mathcal{L}_{\exp}} = \sup_{0 < t < 1} \big(\log(e/t)\big)^{-1} f^{**}(t) < +\infty.$$

Theorem 11.15 *The dual space of $\mathcal{L}\log\mathcal{L}$ is $\mathcal{L}_{\exp}$. There is some constant C such that*

$$\|g\|_{\mathcal{L}_{\exp}} \leq \|g\|_{\mathcal{L}_{\exp}}^{\bullet} \leq C\|g\|_{\mathcal{L}_{\exp}},$$

Where $\|\cdot\|_{\mathcal{L}_{\exp}}^{\bullet}$ denotes the norm on $\mathcal{L}_{\exp}$ as a dual of $\mathcal{L}\log\mathcal{L}$.

Proof. First, assume that $f \in \mathcal{L}\log\mathcal{L}$ and $g \in \mathcal{L}_{\exp}$, then fg is integrable and

$$\left|\int_X fg\,d\mu\right| \leq C\|f\|_{\mathcal{L}\log\mathcal{L}}\|g\|_{\mathcal{L}_{\exp}}. \tag{11.2}$$

To see it, write $f = \sum_{j=1}^{\infty} a_j \chi_{A_j}$ as in Theorem 11.14. Then

$$\left|\int_X fg\,d\mu\right| \leq \sum_{j=1}^{\infty} |a_j| \int_{A_j} |g|\,d\mu \leq \sum_{j=1}^{\infty} |a_j|\mu(A_j)g^{**}(\mu(A_j))$$

$$\leq \sum_{j=1}^{\infty} |a_j|\mu(A_j)\|g\|_{\mathcal{L}_{\exp}} \log(e/\mu(A_j))$$

$$\leq \sum_{j=1}^{\infty} |a_j|\|\chi_{A_j}\|_{\mathcal{L}\log\mathcal{L}}\|g\|_{\mathcal{L}_{\exp}} \leq C\|f\|_{\mathcal{L}\log\mathcal{L}}\|g\|_{\mathcal{L}_{\exp}}.$$

Assume that u is a continuous linear functional on $\mathcal{L}\log\mathcal{L}$. Then we define $\nu(A) = u(\chi_A)$. This is an additive function defined on measurable sets. Now $|\nu(A)| \leq \|u\|\mu(A)\log(e/\mu(A))$. From this inequality it follows that ν is a signed countably additive measure and that $\nu \ll \mu$. Then the Radon-Nikodym Theorem gives us a measurable function g such that for every measurable set A, $u(A) = \int_A g\,d\mu$. This function g is in $\mathcal{L}_{\exp}(\mu)$. Indeed, we know

$$\left|\int_A g\,d\mu\right| \leq \|u\|\mu(A)\log(e/\mu(A)).$$

Then, by Proposition 11.5, we obtain $g^{**}(t) \leq \|u\|\log(e/t)$. That is, $g \in \mathcal{L}_{\exp}$ and $\|g\|_{\mathcal{L}_{\exp}} \leq \|u\|$.

Now for every simple function we have $u(\varphi) = \int \varphi g\,d\mu$. By the density of simple functions in the space $\mathcal{L}\log\mathcal{L}$ (this follows from Theorem 11.14) and (11.2), we derive that

$$u(f) = \int_X fg\,d\mu, \qquad \forall f \in \mathcal{L}\log\mathcal{L}.$$

Then we have proved that $\mathcal{L}_{\exp}$ is the dual space to $\mathcal{L}\log\mathcal{L}$. Besides, we have proved the two inequalities between the norms. $\square$

Finally note the following examples of spaces of functions

Take $\varphi(t) = (\log^+ t)^2$. We obtain the space $\mathcal{L}(\log\mathcal{L})^2$, that is, the set of measurable functions f such that

$$\int_X |f|(\log^+|f|)^2\,d\mu < +\infty.$$

The norm in this case is given by

$$\|f\|_{\mathcal{L}(\log\mathcal{L})^2} = \int_0^1 f^*(t)(\log 1/t)^2\,dt < +\infty.$$

The norm of a characteristic function χ_A with $\mu(A) = m$ is equal to $m(\log e/m)^2 + m$. By the general theory this is also an atomic space

Theorem 11.16 *There is an absolute constant C such that for every function $f \in \mathcal{L}(\log\mathcal{L})^2$ there exist a sequence of measurable sets $(A_j)_{j=1}^\infty$ and a sequence of complex numbers (a_j) such that $f = \sum_{j=1}^\infty a_j\chi_{A_j}$ and*

$$\|f\|_{\mathcal{L}(\log\mathcal{L})^2} \leq \sum_{j=1}^\infty |a_j|\|\chi_A\|_{\mathcal{L}(\log\mathcal{L})^2} \leq C\|f\|_{\mathcal{L}(\log\mathcal{L})^2}.$$

Two more spaces have a role in the theory of pointwise convergence of Fourier series. They are $\mathcal{L}\log\mathcal{L}\log\log\mathcal{L}$ and $\mathcal{L}\log\mathcal{L}\log\log\log\mathcal{L}$. They are defined as the set of measurable functions such that, respectively, the integrals

$$\int_X |f|(\log^+|f|)(\log^+\log^+|f|)\,d\mu$$

and

$$\int_X |f|(\log^+|f|)(\log^+\log^+\log^+|f|)\,d\mu,$$

are finite.

Define $L(t) = L_1(t) = \log(1 + t)$ and for every integer $n > 2$ put $L_n(t) = L(L_{n-1}(t))$. These are well defined functions $L_n\colon [0, +\infty) \to [0, +\infty)$, and it is easily verified, by induction, that $L_n(t^2) \leq 2L_n(t)$. Then if we put $\varphi(t) = L_1(t)L_3(t)$ we will have $\varphi(t^2) \leq 4\varphi(t)$. All the other conditions (2), (3) and (4) are easily verified. It is clear that with this choice of φ we obtain the space $\mathcal{L}\log\mathcal{L}\log\log\log\mathcal{L}$. The other one is analogous.

12. The maximal operator of Fourier series

12.1 Introduction

Let $f: I \to \mathbf{C}$ be a measurable function such that $|f| = \chi_A$. For every $y > 0$ and $1 < p < +\infty$ we have

$$\mathfrak{m}\{x \in I/2 : C_I^* f(x) > y\} \leq \left(\frac{Cp^2}{p-1}\right)^p \frac{\mathfrak{m}(A)}{y^p}. \qquad (12.1)$$

This is the basic result of the Carleson-Hunt construction. In this chapter we turn our attention to the maximal operator S^* of Fourier series. First, in section 2 we obtain an analogous condition to (12.1) for S^*. In section 3 we prove that our knowledge about S^* can be summarized by an inequality for the decreasing rearrangement $(S^* f)^*$ of $S^* f$. Then we derive Hunt's theorems about the continuity of the operator S^* between the spaces $\mathcal{L}^\infty \to \mathcal{L}_{\exp}$; $\mathcal{L}(\log \mathcal{L})^2 \to \mathcal{L}^1$; and $\mathcal{L}^p \to \mathcal{L}^p$ ($1 < p < +\infty$). Finally we define two quasi-Banach spaces Q and QA, and prove that S^* maps them to $\mathcal{L}^{1,\infty}$. These results contain those of Sjölin, Soria, and Antonov.

12.2 Maximal operator of Fourier series

All the considerations of this chapter can be applied to the operator C_I^*, but we prefer to talk about the maximal operator of Fourier series

$$S^* f(x) = \sup_{n \in \mathbf{N}} |S_n(f, x)|.$$

First, we prove that $S^* f$ satisfies the same restricted weak inequality.

In this chapter the functions usually are defined on $[-\pi, \pi]$ and we denote by μ the normalized Lebesgue measure on this interval.

Proposition 12.1 *There exists an absolute constant $C < +\infty$ such that for every measurable function $f: [-\pi, \pi] \to \mathbf{R}$ with $|f| = \chi_A$, every $1 < p < +\infty$ and $y > 0$ we have*

$$\mu\{x \in [-\pi, \pi] : S^* f(x) > y\} \leq C_p^p \frac{\mu(A)}{y^p}, \qquad C_p \leq C \frac{p^2}{p-1}. \qquad (12.2)$$

Proof. Let $f^\circ: \mathbf{R} \to \mathbf{R}$ be equal to 0 for $|t| > 2\pi$ and equal to the periodic extension of f on $[-2\pi, 2\pi]$.

Recalling the expression of the Dirichlet kernel (2.3), we obtain, for every $x \in [-\pi, \pi]$,

$$S_n(f, x) = \frac{1}{\pi} \int_{-\pi}^{\pi} f^\circ(x - t) \frac{\sin nt}{t} \, dt + \frac{1}{2\pi} \int_{-\pi}^{\pi} f^\circ(x - t) \varphi_n(t) \, dt.$$

Since $\sin nt/t$ is uniformly bounded in $[\pi, 3\pi]$ and $[-3\pi, -\pi]$, and $\|\varphi_n\|_\infty$ is uniformly bounded in $n \in \mathbf{N}$, we get

$$|S_n(f, x)| \leq \frac{1}{\pi} \left| \int_{-2\pi}^{2\pi} f^\circ(t) \frac{\sin n(x - t)}{x - t} \, dt \right| + C\mathfrak{m}(A), \qquad |x| < \pi.$$

Therefore if I denotes the interval $[-2\pi, 2\pi]$, we obtain

$$S^* f(x) \leq C_I^* f^\circ(x) + C\mathfrak{m}(A).$$

Since $|f^\circ| = \chi_{A'}$ with $\mathfrak{m}(A') \leq 2\mathfrak{m}(A)$, for $y > 2C\mathfrak{m}(A)$ we have

$$\mathfrak{m}\{S^* f(x) > y\} \leq \mathfrak{m}\{C_I^* f^\circ(x) > y/2\} \leq (2C_p)^p \frac{2\mathfrak{m}(A)}{y^p},$$

where $C_p \leq Cp^2/(p-1)$ is the constant in (12.1).

Finally we arrive at

$$\mathfrak{m}\{S^* f(x) > y\} \leq D_p^p \frac{\mathfrak{m}(A)}{y^p}, \qquad D_p \leq D\frac{p^2}{p-1}.$$

Taking D big enough we can eliminate the restriction $y > 2C\mathfrak{m}(A)$. Finally we change to the measure μ. $\qquad\square$

Now we can prove another extension of this result.

Theorem 12.2 *There is an absolute constant C such that given a measurable function f on $[-\pi, \pi]$ with $\|f\|_\infty = 1$ and a measurable set A such that $f(x) = 0$ for every $x \notin A$, then for every $1 < p < +\infty$ and $y > 0$ we have*

$$\mu\{S^* f > y\} \leq \left(\frac{C_p}{y}\right)^p \mu(A), \qquad C_p \leq Cp^2/(p-1).$$

Proof. First assume f to be real. Then we can write

$$f = \Re(e^{ig(x)} \chi_A(x)),$$

where $g(x) = \arccos f(x)$ is conveniently defined so as to be measurable. Therefore we have $f = (f_1 + f_2)/2$ with f_1 and f_2 being special functions in our meaning, that is $|f_1| = |f_2| = \chi_A$.

Then we have $S^* f \leq (S^* f_1 + S^* f_2)/2$. Therefore for every $y > 0$,

$$\{S^*f > y\} \subset \{S^*f_1 > y\} \cup \{S^*f_2 > y\}.$$

Hence, by the basic result, we derive

$$\mu\{S^*f > y\} \leq 2\left(\frac{C_p}{y}\right)^p \mu(A) \leq \left(\frac{2C_p}{y}\right)^p \mu(A).$$

For a complex f the theorem follows easily from the real case.   □

12.3 The distribution function of S^*f

Our first objective is to obtain the information contained in these inequalities about the distribution function of S^*f. We shall need a function φ_a to describe this distribution function

$$\varphi_a(t) = \begin{cases} (1/t)\log(e/t) & \text{if } 0 < t < 1, \\ e^{a(1-t)} & \text{if } 1 \leq t < +\infty. \end{cases}$$

Lemma 12.3 *(a) Given $A > 0$ there exist N and $a > 0$ such that for every $y > 0$ there exists $p > 1$ with*

$$\frac{1}{y^p}\left(\frac{Ap^2}{p-1}\right)^p \leq N\varphi_a(y).$$

(b) Given M and $a > 0$, there is a constant $A > 0$ such that for every $y > 0$ and every $p > 1$

$$M\varphi_a(y) \leq \frac{1}{y^p}\left(\frac{Ap^2}{p-1}\right)^p.$$

Proof. (a) Given A, take $p = \sup(\alpha y, 2)$ when $y > 1$, with $\alpha > 0$ conveniently chosen; and $p = 1 + \left(\log(e/y)\right)^{-1}$ when $y < 1$.

(b) For $y > 1$ the inequality is equivalent to $(Me^a)^{1/p}ye^{-ay/p} \leq Ap^2/(p-1)$. For a fixed p the function $ye^{-ay/p} \leq p/ae$. Therefore we must prove $(Me^a)^{1/p} \leq Aaep/(p-1)$. Now it is easy to see that this is true for every value of $p > 1$ if we choose a convenient A.

For $y < 1$ we have to prove $My^{p-1}\log(e/y) \leq \left(Ap^2/(p-1)\right)^p$. For a fixed p, $My^{p-1}\log(e/y)$ attains a maximum at $y = e\exp(-(p-1)^{-1})$ and for this value the inequality can be easily checked.   □

From Theorem 12.2 and Lemma 12.3 we obtain the following theorem. This is equivalent to Theorem 12.2 and contains all the information that the construction of Carleson-Hunt gives about the operator S^*.

Theorem 12.4 *There exist absolute constants a, and B such that if f is a measurable function with $\|f\|_\infty \leq 1$ and A a measurable set such that $f(x) = 0$ for every $x \notin A$, then for every $y > 0$,*

$$\mu\{S^*f > y\} \leq B\varphi_a(y)\mu(A).$$

12.4 The operator S^*f on the space $\mathcal{L}_\infty$

Theorem 12.5 (Hunt) *There are absolute constants A and $B > 0$ such that for every $f \in \mathcal{L}^\infty[-\pi, \pi]$, and every $y > 0$*

$$\mu\{S^*f(x) > y\} \leq Ae^{-By/\|f\|_\infty}.$$

Proof. By homogeneity we can assume that $\|f\|_\infty = 1$. If we take A such that $Ae^{-B} > 1$ the inequality says something only for $y > 1$. In the case $y > 1$ the inequality is the same as that of Theorem 12.4. $\qquad\Box$

The set of measurable functions g that satisfies an inequality of the type $\mu\{|g| > y\} \leq Ae^{-By}$, forms a Banach space. This space coincides with the space of functions for which there is a constant C such that $\int e^{C|g|}\,d\mu < +\infty$, and is usually called $\mathcal{L}_{\exp}$. As we have seen in chapter eleven, a norm on this space is given by

$$\|g\|_{\mathcal{L}_{\exp}} = \sup_{0<t<1} \frac{g^{**}(t)}{\log(e/t)}.$$

Here g^{**} is the maximal function defined as

$$g^{**}(t) = \sup\left\{\frac{1}{\mu(E)}\int_E g\,d\mu : \mu(E) = t\right\}.$$

If we denote by g^* the decreasing rearrangement of g, then we have

$$g^{**}(t) = \frac{1}{t}\int_0^t g^*(t)\,dt.$$

Since $S^*(f+g) \leq S^*f + S^*g$ it follows that $|S^*(f+h) - S^*(f)| \leq S^*(h)$. Therefore the theorem implies that $S^*: \mathcal{L}^\infty \to \mathcal{L}_{\exp}$ is continuous. Thus there is a constant C such that $\|S^*f\|_{\mathcal{L}_{\exp}} \leq C\|f\|_\infty$.

12.5 The operator S^* on the space $\mathcal{L}(\log \mathcal{L})^2$

We derive from Theorem 12.4 that the operator S^* maps the space $\mathcal{L}(\log \mathcal{L})^2$ into $\mathcal{L}^1$. First, the space $\mathcal{L}(\log \mathcal{L})^2$ is the set of measurable functions g such that $|g|(\log^+ |g|)^2$ is integrable. This is a Banach space. We have defined its norm as

$$\|g\|_{\mathcal{L}(\log \mathcal{L})^2} = \int_0^1 g^*(t)(\log t)^2 \, dt.$$

Theorem 12.6 (Hunt) *The operator $S^*: \mathcal{L}(\log \mathcal{L})^2 \to \mathcal{L}^1$ is continuous. Hence there is a constant C such that for every measurable function f*

$$\|S^* f\|_1 \leq C \|f\|_{\mathcal{L}(\log \mathcal{L})^2}.$$

Proof. Given f there is a sequence of pairwise disjoint measurable sets $(A_j)_{j=1}^\infty$ such that $\mu(A_j) = 2^{-j}$, and for every $j < k$, $x \in A_j$ and $y \in A_k$ we have $|f(x)| \leq |f(y)|$. Put $a_j = \sup\{|f(x)| : x \in A_j\}$. Then if $f_j = f \chi_{A_j}$, we have that $f = \sum_{j=1}^\infty f_j$ and $\|f_j\|_\infty \leq a_j$. We deduce easily the following inequality for the decreasing rearrangement:

$$f^*(t) \geq \sum_{j=1}^\infty a_j \, \chi_{[2^{-j-1}, 2^{-j})}$$

Now by Theorem 12.4 for $f \in \mathcal{L}^\infty$, supported on the measurable set A of measure $\mu(A) = m$, and bounded by 1, we have

$$\|S^* f\|_1 = \int_0^{+\infty} \mu\{S^* f > y\} \, dy \leq \int_0^m dy + \int_m^\infty B m \varphi_a(y) \, dy$$
$$\leq C m (\log e/m)^2.$$

Therefore by the subadditivity of S^* we get, for a general f,

$$\|S^* f\|_1 \leq C \sum_{j=1}^\infty a_j \frac{1}{2^j} (\log e2^j)^2.$$

Thus

$$\|S^* f\|_1 \leq C' \sum_{j=1}^\infty a_j \frac{1}{2^{j+1}} (\log 2^{(j+1)})^2 \leq C \int_0^1 f^*(t)(\log t)^2 \, dt.$$

$\square$

12.6 The operator S^* on the space $\mathcal{L}^p$

Here we derive the principal result of Hunt. Observe that our notation for the norms of $\mathcal{L}^{p,1}(\mu)$ and $\mathcal{L}^{p,\infty}(\mu)$ is not standard. We have put, for $1 < p < +\infty$,

$$\|f\|_{p,1} = \frac{1}{p}\int_0^1 t^{1/p}f^*(t)\frac{dt}{t}, \qquad \|f\|_{p,\infty} = \sup_{0<t} t^{1/p}f^{**}(t).$$

They are specially useful for our purpose here.

Theorem 12.7 *For every $p \in (1,+\infty)$, the operator S^* maps $\mathcal{L}^{p,1}(\mu)$ into $\mathcal{L}^{p,\infty}(\mu)$, and*

$$\|S^*f\|_{p,+\infty} \leq C\frac{p^3}{(p-1)^2}\|f\|_{p,1}.$$

Proof. Proposition 12.2 gives us, for every characteristic function,

$$\sup_{y>0} y\big(\mu\{S^*\chi_A > y\}\big)^{1/p} \leq \frac{Cp^2}{p-1}\|\chi_A\|_p.$$

For every function f, the decreasing rearrangement f^* and the distribution function $\mu_f(y) = \mu\{f > y\}$ are essentially inverse functions, therefore

$$\sup_{0<t<1} t^{1/p}f^*(t) = \sup_{y>0} y\mu_f(y)^{1/p}.$$

Thus, for every characteristic function,

$$\sup_{0<t<1} t^{1/p}(S\chi_A)^*(t) \leq C\frac{p^2}{p-1}\|\chi_A\|_p.$$

We have seen in (11.1) that

$$\sup_{0<t<1} t^{1/p}f^{**}(t) \leq \frac{p}{p-1}\sup_{0<t<1} t^{1/p}f^*(t).$$

Thus we have

$$\|S^*\chi_A\|_{p,\infty} \leq \frac{Cp^3}{(p-1)^2}\|\chi_A\|_p.$$

Now, recall that $\mathcal{L}^{p,1}(\mu)$ is an atomic space. That is, each function $f \in \mathcal{L}^{p,1}(\mu)$ can be written as $f = \sum_{j=1}^\infty a_j\chi_{A_j}$ in such a way that $\sum_{j=1}^\infty |a_j|\|\chi_{A_j}\|_p \leq C\|f\|_{p,1}$.

Hence

$$S^*f \leq \sum_{j=1}^\infty |a_j|S^*(\chi_{A_j}).$$

Therefore

$$\|S^*f\|_{p,\infty} \leq \sum_{j=1}^{\infty} |a_j| \|S^*(\chi_{A_j})\|_{p,\infty} \leq C\frac{p^3}{(p-1)^2} \sum_{j=1}^{\infty} |a_j| \|\chi_{A_j}\|_p$$

$$\leq C'\frac{p^3}{(p-1)^2}\|f\|_{p,1}.$$

$\square$

Theorem 12.8 (Carleson-Hunt) *For every $1 < p < +\infty$, the operator $S^*\colon \mathcal{L}^p \to \mathcal{L}^p$ is continuous. More precisely, there is a constant C such that for every measurable function f*

$$\|S^*f\|_p \leq C\frac{p^4}{(p-1)^3}\|f\|_p.$$

Therefore for $f \in \mathcal{L}^p[-\pi, \pi]$

$$\lim_{n\to\infty} S_n(f,x) = f(x), \qquad \text{a. e. on } [-\pi, \pi].$$

Proof. By the previous theorem we know that

$$\|S^*f\|_{p,+\infty} \leq C\frac{p^3}{(p-1)^2}\|f\|_{p,1},$$

for every $1 < p < +\infty$. Thus we can apply Marcinkiewicz's Theorem 11.9. Therefore if $p_1 < p < p_0$, the norm of the operator $S^*\colon \mathcal{L}^p(\mu) \to \mathcal{L}^p(\mu)$ is bounded by

$$\|S^*\|_p \leq C\frac{p(p_0-p_1)}{(p_0-p)(p-p_1)}\left(\frac{p_0^3}{(p_0-1)^2}\right)^{1-\theta}\left(\frac{p_1^3}{(p_1-1)^2}\right)^{\theta}.$$

Observe that if p_0' and p_1' denote the conjugate exponents, then $p_0^3/(p_0-1)^2 = p_0 p_0'^2$. Thus if we take p_0 and p_1 conjugate exponents,

$$\|S^*\|_p \leq C\frac{p(p_0-p_1)}{(p_0-p)(p-p_1)}p_0^{1+\theta}p_1^{2-\theta}.$$

In the case $1 < p < 2$ we choose $p_0 = (p+1)/(p-1)$ and $p_1 = (p+1)/2$. With these choices p_0 and p_1 are conjugates, $\theta = (1+2p-p^2)/p(3-p)$, and we get

$$\|S^*\|_p \leq C\frac{2^\theta p(p+1)^4(3-p)}{4(1+2p-p^2)(p-1)^{2+\theta}} \leq \frac{C'}{(p-1)^3}.$$

In the case $2 < p < +\infty$ we choose $p_0 = 2p$, $p_1 = 2p/(2p-1)$, then $\theta = 1/2(p-1)$ and we get

$$\|S^*\|_p \leq C\frac{2(p-1)(2p)^4}{p(2p-3)(2p-1)^{2-\theta}} \leq C'p.$$

Therefore for all values of p we arrive at

$$\|S^*\|_p \leq C \frac{p^4}{(p-1)^3}.$$

The conclusion about the pointwise convergence is now standard. We know that for a dense set of functions on $\mathcal{L}^p[-\pi, \pi]$ we have pointwise convergence at all points.

Now we introduce the operator

$$\Omega f(x) = \limsup_n |S_n(f, x) - f(x)|.$$

It is clear that $\Omega f(x) \leq S^* f(x) + |f(x)|$. Also $\Omega(f) = \Omega(f - \varphi)$, for every φ periodic and differentiable. Then for every positive real number α and φ as before

$$\{\Omega f(x) > 2\alpha\} \subset \{S^*(f - \varphi) > \alpha\} \cup \{|f(x) - \varphi(x)| > \alpha\},$$

from which we deduce easily that $\{\Omega f(x) > 2\alpha\}$ has measure zero.   □

12.7 The maximal space Q

We are interested here in a space of functions with almost everywhere convergent Fourier series. First, observe the following consequence of Theorem 12.4

Corollary 12.9 *There exists an absolute constant C, such that if f is a measurable function with $\|f\|_\infty \leq 1$ and A is a measurable set with $\mu(A) = m$ and such that $f(x) = 0$ for every $x \notin A$, then for every $y > 0$ we have*

$$\mu\{S^* f > y\} \leq C \frac{m \log(e/m)}{y}.$$

Proof. By corollary 12.4 we know that $\mu\{S^* f > y\} \leq Bm\varphi_a(y)$. Since μ is a probability, $\mu\{S^* f > y\} \leq 1$. Therefore, we have to prove that there exists C such that $\inf(1, Bm\varphi(y)) \leq Cm \log(e/m)/y$. This is easily checked.   □

In 1961 Stein [44, p. 169] proved that if X is a Banach space of functions defined on $[-\pi, \pi]$ and such that $\|f\|_1 \leq \|f\|_X$, and for every $f \in X$, $S^* f(x) < +\infty$ for x in a set of positive measure, then $\mu\{S^* f > y\} \leq Ay^{-1}\|f\|_X$ and in case that $X = \mathcal{L}^p$ ($1 < p < 2$), then $\mu\{S^* f > y\} \leq Ay^{-p}\|f\|_p$. This made Zygmund think that the conjecture of Luzin must be false. Since the construction of Carleson-Hunt gives $\mu\{S^* f > y\} \leq Ay^{-1}\|f\|_{\mathcal{L} \log \mathcal{L}}$ for characteristic functions, the best we could hope to prove is a. e. convergence for functions in the space $\mathcal{L}\log\mathcal{L}$. But this appears very difficult to achieve.

We have here some functions f with $S^* f \in \mathcal{L}^{1,\infty}$, therefore we are interested in conditions of stability of the space $\mathcal{L}^{1,\infty}$.

The principal result in this direction is the following theorem of Stein and Weiss (but we give a proof due to Kalton).

For typographical reasons we put $\|f\|_1^*$ instead of $\|f\|_{1,\infty}$

Theorem 12.10 (Stein–Weiss) *Let (f_j) be a sequence of functions on $\mathcal{L}^{1,\infty}$, with $\sum_{j=1}^{\infty} \|f_j\|_1^* = 1$. If $\sum_{j=1}^{\infty} \|f_j\|_1^* \log(e/\|f_j\|_1^*) < +\infty$, the sum $\sum_{j=1}^{\infty} f_j$ is almost everywhere absolutely convergent and*

$$\left\| \sum_{j=1}^{\infty} f_j \right\|_1^* \le 4 \sum_{j=1}^{\infty} \|f_j\|_1^* \log \frac{e}{\|f_j\|_1^*}.$$

Proof (Kalton). Let A be a measurable set of measure $\mu(A) = m$. Define the sets $B_j = \{|f_j| > 2/m\}$, and $B = \bigcup_j B_j$. We have

$$\mu(B) \le \sum_{j=1}^{\infty} \mu(B_j) \le \sum_{j=1}^{\infty} \frac{1}{2}\|f_j\|_1^* m \le m/2.$$

Then

$$\inf_{t \in A} |f(t)| \le \inf_{t \in A \setminus B} |f(t)| \le \frac{2}{m} \int_{A \setminus B} |f|\,d\mu \le \frac{2}{m} \sum_{j=1}^{\infty} \int_{A \setminus B} |f_j|\,d\mu$$

$$\le \frac{2}{m} \sum_{j=1}^{\infty} \int_{A \setminus B_j} |f_j|\,d\mu \le \frac{2}{m} \sum_{j=1}^{\infty} \int_A \inf(|f_j|, 2/m)\,d\mu$$

$$\le \frac{2}{m} \sum_{j=1}^{\infty} \int_0^{2/m} \inf(\|f_j\|_1^*/t, m)\,dt = \frac{2}{m} \sum_{j=1}^{\infty} \{\|f_j\|_1^* + \|f_j\|_1^* \log 2/\|f_j\|_1^*\}$$

$$\le \frac{2}{m}\{1 + \sum_{j=1}^{\infty} \|f_j\|_1^* \log e/\|f_j\|_1^*\} \le \frac{4}{m} \sum_{j=1}^{\infty} \|f_j\|_1^* \log e/\|f_j\|_1^*.$$

Therefore if this quantity is $4L/m$ we have

$$\mu\{|f| > 4L/m\} < m,$$

because on every set A of measure m we have some points where $|f|$ is less than $4L/m$. This is equivalent to saying that $\mu\{|f| > y\} \le 4L/y$. $\square$

The quantity that appears in the above theorem is not homogeneous, and we have proved the inequality only when $\sum_{j=1}^{\infty} \|f_j\|_1^* = 1$. These difficulties can be avoided.

Theorem 12.11 (Kalton) *For every sequence* (f_j) *on* $\mathcal{L}^{1,\infty}$ *we have*

$$\left\|\sum_{j=1}^{\infty} f_j\right\|_1^* \leq C \sum_{j=1}^{\infty} (1 + \log j)\|f_j\|_1^*.$$

(C is an absolute constant).

Proof. Obviously we can assume that $\sum_{j=1}^{\infty} \|f_j\|_1^* = 1$. So, by the above theorem we only need to prove that there is an absolute constant such that if $a_j > 0$ and $\sum_{j=1}^{\infty} a_j = 1$, then

$$\sum_{j=1}^{\infty} a_j \log e/a_j \leq C \sum_{j=1}^{\infty} a_j^*(1 + \log j),$$

where (a_j^*) denotes the decreasing rearrangement of the sequence (a_j).

Observe that $\sum_{j=1}^{\infty} a_j^*(1 + \log j)$ is the infimum of all possible sums $\sum_{j=1}^{\infty} a_{\sigma(j)}(1 + \log j)$ for every permutation σ of the natural numbers

Hence we have to prove the following assertion:

For every $\alpha > 1$, the supremum of the function

$$\sum_{j=1}^{N} x_j \log(e/x_j) - \alpha \sum_{j=1}^{N} x_j(1 + \log j)$$

in the set of $(x_1, x_2, \ldots, x_N)$ with $x_1, x_2, \ldots x_N \geq 0$ and $\sum_{j=1}^{N} x_j = 1$ is equal to

$$\log\left(e \sum_{j=1}^{N} \frac{1}{j^\alpha}\right) - \alpha \tag{12.3}$$

We proceed by induction on N. For $N = 1, 2$ the result can be easily verified. Also observe that, for every dimension, there is a critical point, interior to the region we are considering. This point can be obtained by Lagrange multipliers. The computations give $a_k = k^{-\alpha}\left(\sum_{j=1}^{N} j^{-\alpha}\right)^{-1}$. From the fact we see that the function attains the value (12.3).

We assume the result to be true in the case $N - 1$. In the case of N variables, the maximum of the function exists because it is a continuous function on a compact set. It is clear that the maximum is attained at a point where $x_1 \geq x_2 \geq \cdots \geq x_N \geq 0$. So we assume that for $x_j = a_j$ the function takes the maximum value. Every $a_j > 0$, because if $a_j = 0$ for some j, then $a_N = 0$ and the maximum in this case would be the same as in that with $N-1$ variables. This contradicts the fact that we have obtained a greater value at an interior point. Therefore the maximum is a relative maximum and must coincide with the point calculated.

Therefore, taking some concrete value for $\alpha > 1$ we obtain for $x_j \geq 0$ and $\sum_{j=1}^{\infty} x_j = 1$

$$\sum_{j=1}^{\infty} x_j \log e/x_j \leq \log(e\zeta(\alpha)) + \alpha \sum_{j=1}^{\infty} x_j (1 + \log j) \leq C \sum_{j=1}^{\infty} x_j (1 + \log j).$$

$\square$

Remark. Observe that from $\sum_{j=1}^{\infty} a_j = 1$ it follows that $a_j^* j \leq 1$, thus

$$\sum_{j=1}^{\infty} a_j^* \log(e/a_j^*) \geq \sum_{j=1}^{\infty} a_j^* (1 + \log j).$$

Therefore Theorem 12.11 is equivalent to Stein-Weiss theorem.

A **quasi-norm** on a real or complex vector space X is a map $x \mapsto \|x\|$ such that $\|x\| \geq 0$ and $\|x\| = 0$ if and only if $x = 0$, $\|ax\| = |a|\,\|x\|$ for every scalar a and vector x, and finally there is some constant $C > 0$ such that $\|x + y\| \leq C(\|x\| + \|y\|)$. A quasi norm induces a compatible topology. If this space is complete it is called a **quasi-Banach space**. $\mathcal{L}^{1,\infty}$ is a quasi-Banach space.

A quasi-Banach space is called **logconvex** if it satisfies the conclusion of Proposition 12.11.

We define now the space Q of all measurable functions f such that there exist a sequence (a_j) of positive real numbers and measurable sets A_j of measure m_j, such that

$$|f| \leq \sum_{j=1}^{\infty} a_j \chi_{A_j}, \qquad \sum_{j=1}^{\infty} a_j m_j \log(e/m_j)(1 + \log j) < +\infty.$$

And for every such function we define

$$\|f\|_Q = \inf\left\{ \sum_{j=1}^{\infty} a_j m_j \log(e/m_j)(1 + \log j) \right\},$$

where we consider all possible sequences (a_j) and (A_j).

Proposition 12.12 Q *is a logconvex quasi-Banach space.*

Proof. From the definition it is clear that for $f \in Q$, $\|f\|_Q \geq 0$. To see that the quasi-norm vanishes only for $f \sim 0$, we obtain the inequality $\|f\|_{\mathcal{L}\log\mathcal{L}} \leq \|f\|_Q$. This is true because from $|f| \leq \sum_{j=1}^{\infty} a_j \chi_{A_j}$ we derive

$$\|f\|_{\mathcal{L}\log\mathcal{L}} \leq \sum_{j=1}^{\infty} a_j \|\chi_{A_j}\|_{\mathcal{L}\log\mathcal{L}} = \sum_{j=1}^{\infty} a_j m_j \log(e/m_j)$$

$$\leq \sum_{j=1}^{\infty} a_j m_j \log(e/m_j)(1 + \log j).$$

The equality $\|af\|_Q = |a|\,\|f\|_Q$ is a simple consequence of the definition.

Assume now that we have $f = \sum_{j=1}^{\infty} f_j$. Given $\varepsilon > 0$, we determine positive real numbers a_{jk} and measurable sets A_{jk} such that

$$\|f_j\|_Q \leq \sum_{k=1}^{\infty} a_{jk} m_{jk} \log(e/m_{jk})(1 + \log k) \leq \|f_j\|_Q + \frac{\varepsilon}{2^j}(1 + \log j)^{-1}.$$

It follows that

$$\|f\|_Q \leq \sum_{jk} a_{jk} m_{jk} \log(e/m_{jk})(1 + \log \sigma(j,k))$$

where $\sigma \colon \mathbf{N} \times \mathbf{N} \to \mathbf{N}$ is a bijective application. For example, take the usual bijection $\sigma(j,k) = j + (j+k-2)(j+k-1)/2$. Then we have $(1 + \log \sigma(j,k)) \leq 2(1 + \log j)(1 + \log k)$ Therefore

$$\|f\|_Q \leq \varepsilon + 2\sum_{j=1}^{\infty}(1 + \log j)\|f_j\|_Q$$

This implies that $\|\cdot\|_Q$ is a logconvex quasi-norm.

To prove that the space is complete we only need to prove that if $\|f_j\|_Q \leq 2^{-j}$ the series $\sum_{j=1}^{\infty} f_j$ is convergent in Q. This is obvious from the definition and the logconvexity of the norm. $\qquad\square$

Proposition 12.13 *For every measurable f*

$$\mu\{S^* f > y\} \leq C\frac{\|f\|_Q}{y},$$

where $C < +\infty$ is an absolute constant. Therefore the Fourier series of every function in Q is pointwise almost everywhere convergent.

Proof. If $|f| \leq \sum_{j=1}^{\infty} a_j \chi_{A_j}$, then there exist measurable functions f_j with $\|f_j\|_{\infty} \leq 1$ and supported on A_j, and such that $f = \sum_{j=1}^{\infty} a_j f_j$.

It follows from Corollary 12.9 that $\|S^* f_j\|_1^* \leq C m_j \log(e/m_j)$, where $m_j = \mu(A_j)$. Therefore from Kalton theorem

$$\|S^* f\|_1^* \leq C\sum_{j=1}^{\infty} a_j(1 + \log j)\|S^* f_j\|_1^* \leq C'\sum_{j=1}^{\infty}(1 + \log j)a_j m_j \log(e/m_j).$$

Taking the infimum for all possible decompositions we arrive at $\|S^*f\|_1^* \leq C'\|f\|_Q$.
□

Remark. It is easy to see that the space $B_{\varphi_1}^*$ defined by Soria [42] is contained in Q. But it is not clear whether $Q = B_{\varphi_1}^*$.

Remark. We can give an example of a sublinear operator $T: Q \to \mathcal{L}^{1,\infty}$ such that for every $f \notin Q$, we have $\sup\{Tg : g \in Q, |g| \leq |f|\} = +\infty$. But this operator satisfies the conclusion of Corollary 12.9. Therefore in this sense the space Q is maximal. But obviously it may be possible that we can derive better properties for S^* assuming only the properties given in Theorem 12.4. As we will see in the following section, it is also possible to obtain some extra information for the special operator S^* in which we are interested.

12.8 The theorem of Antonov

Now we are going to apply the ideas of Antonov to obtain a quasi-Banach space $QA \supset Q$ in which we can prove a. e. convergence. In particular we shall see that $QA \supset \mathcal{L}\log\mathcal{L}\log\log\log\mathcal{L}$. We shall need some information about the Dirichlet kernel that is due to Antonov.

Proposition 12.14 (Antonov) *Let $f: [-\pi, \pi] \to \mathbf{R}$ be a measurable function and A a measurable set such that $0 \leq f(x) \leq a$ for every $x \in A$ and $f(x) = 0$ for every $x \notin A$. Then for every $\varepsilon > 0$ and $n \in \mathbf{N}$, there is a measurable set $B \subset A$ such that $a\mu(B) = \|f\|_1$ and such that*

$$|S_j(f, x) - S_j(a\chi_B, x)| \leq \varepsilon, \quad \text{for every } x \in [-\pi, \pi], \text{ and } 1 \leq j \leq n.$$

Proof. Without loss of generality we can assume that $\|f\|_1 > 0$. We are going to divide the interval $[-\pi, \pi]$ into N equal subintervals, where N is big enough. Let $(J_k)_{k=1}^N$ be these intervals. For every such interval we determine a measurable set $B_k \subset J_k \cap A$ such that

$$\int_{J_k \cap A} f \, d\mu = a\mu(B_k).$$

This can be done because $a^{-1} \int_{J_k \cap A} f \, d\mu \leq \mu(J_k \cap A)$. Then we take $B = \bigcup_{k=1}^N B_k$.

Now it is clear that $B \subset A$. Also,

$$a\mu(B) = a\sum_{j=1}^N \mu(B_k) = \sum_{j=1}^N \int_{J_k \cap A} f \, d\mu = \int_A f \, d\mu = \|f\|_1.$$

Now assume that $x \in [-\pi, \pi]$ and $1 \leq j \leq n$. We have

$$|S_j(f, x) - S_j(a\chi_B, x)| = \left| \int_{-\pi}^{\pi} D_j(x - t)(f(t) - a\chi_B(t)) \, d\mu(t) \right|$$

$$\leq \sum_{k=1}^{N} \left| \int_{J_k} D_j(x - t)(f(t) - a\chi_B(t)) \, d\mu(t) \right|$$

$$= \sum_{k=1}^{N} \left| \int_{J_k \cap A} D_j(x - t)(f(t) - a\chi_{B_k}(t)) \, d\mu(t) \right|.$$

Now we notice that $f(t) - a\chi_{B_k}(t)$ has an integral equal to 0 on the set $J_k \cap A$. Therefore if t_k is one extreme of the interval J_k we have

$$|S_j(f, x) - S_j(a\chi_B, x)|$$

$$\leq \sum_{k=1}^{N} \left| \int_{J_k \cap A} (D_j(x - t) - D_j(x - t_k))(f(t) - a\chi_{B_k}(t)) \, d\mu(t) \right|.$$

We apply the mean value theorem, for $t \in J_k$

$$|D_j(x - t) - D_j(x - t_k)| \leq \frac{2\pi}{N} j(j + 1).$$

Therefore

$$|S_j(f, x) - S_j(a\chi_B, x)| \leq \sum_{k=1}^{N} \frac{2\pi}{N} j(j + 1) \int_{J_k \cap A} (f + a\chi_{B_k}) \, d\mu$$

$$\leq \frac{2\pi}{N} j(j + 1) 2\|f\|_1 \leq \varepsilon,$$

if we take

$$N \geq \frac{4\pi n(n + 1)}{\varepsilon} \|f\|_1.$$

$\square$

The procedure of Antonov invites us to define a quasi-norm in the following way.

Definition 12.15 We say that a measurable function f is in QA if the following quantity is finite:

$$\|f\|_{QA} = \inf \left\{ \sum_{j=1}^{\infty} (1 + \log j) \|f_j\|_1 \log \left(\frac{e\|f_j\|_\infty}{\|f_j\|_1} \right) : |f| \leq \sum_{j=1}^{\infty} f_j, \quad f_j \geq 0 \right\}$$

It is clear that for every scalar a we have $\|af\|_{QA} = |a|\|f\|_{QA}$. Furthermore, as in the case of $\|\cdot\|_Q$ we can prove easily that it is a logconvex quasi-norm:

$$\Big\|\sum_{j=1}^{\infty} f_j \Big\|_{QA} \leq \sum_{j=1}^{\infty}(1 + \log j)\|f_j\|_{QA}.$$

With this definition we can prove the following theorem.

Theorem 12.16 *For every measurable f, and $y > 0$*

$$\mu\{S^* f > y\} \leq C\frac{\|f\|_{QA}}{y},$$

where $C < +\infty$ is an absolute constant. Therefore the Fourier series of every function in QA is pointwise almost everywhere convergent.

Proof. To prove the inequality we can assume that $f \in QA$. Every $f \in QA$ is a linear combination of four positive measurable functions $f_j \in QA$, $f = (f_1 - f_2) + i(f_3 - f_4)$ and with $\|f_j\|_{QA} \leq \|f\|_{QA}$. Therefore we can assume that $f \geq 0$.

Then, given $\varepsilon > 0$, we obtain measurable functions $f_j \in \mathcal{L}^{\infty}(\mu)$ such that

$$\|f\|_{QA} \leq \sum_{j=1}^{\infty}(1 + \log j)\|f_j\|_1 \log\Big(\frac{e\|f_j\|_{\infty}}{\|f_j\|_1}\Big) \leq \|f\|_{QA} + \varepsilon.$$

We can assume that $f = \sum_j f_j$ because given $0 \leq h \leq g$ we have

$$\|h\|_1 \log\Big(\frac{e\|h\|_{\infty}}{\|h\|_1}\Big) \leq \|h\|_1 \log\Big(\frac{e\|g\|_{\infty}}{\|h\|_1}\Big) \leq \|g\|_1 \log\Big(\frac{e\|g\|_{\infty}}{\|g\|_1}\Big).$$

Observe that $x \log ea/x$ is increasing on $[0, a))$.

Fix a natural number n. We apply Proposition 12.14 to obtain for every $j \in \mathbf{N}$ a measurable set B_j such that for every $j \in \mathbf{N}$ and $1 \leq k \leq n$ we have

$$\mu(B_j) = \frac{\|f_j\|_1}{\|f_j\|_{\infty}}, \quad |S_k(f_j, x) - S_k(\|f_j\|_{\infty}\chi_{B_j}, x)| \leq \frac{\varepsilon}{2^j}.$$

Now consider the function $g = \sum_{j=1}^{\infty}\|f_j\|_{\infty}\chi_{B_j}$. Its Q-norm is bounded:

$$\|g\|_Q \leq \sum_{j=1}^{\infty}(1 + \log j)\|f_j\|_{\infty}\mu(B_j)\log(e/\mu(B_j))$$

$$= \sum_{j=1}^{\infty}(1 + \log j)\|f_j\|_1 \log\Big(\frac{e\|f_j\|_{\infty}}{\|f_j\|_1}\Big).$$

Therefore

$$\|g\|_Q \leq \|f\|_{QA} + \varepsilon.$$

On the other hand, for $1 \leq k \leq n$

$$|S_k(f, x) - S_k(g, x)| \leq \sum_{j=1}^{\infty} |S_k(f_j, x) - S_k(\|f_j\|_\infty \chi_{B_j}, x)| \leq \varepsilon.$$

Therefore by Proposition 12.13

$$\mu\{\sup_{1 \leq k \leq n} |S_k(f, x)| > y\} \leq \mu\{S^* g > y - \varepsilon\} \leq C \frac{\|g\|_Q}{y - \varepsilon} \leq C \frac{\|f\|_{QA} + \varepsilon}{y - \varepsilon}.$$

Since this inequality is true for every $n \in \mathbf{N}$ we arrive at

$$\mu\{S^* f > y\} \leq C \frac{\|f\|_{QA} + \varepsilon}{y - \varepsilon}.$$

Now this is true for every $\varepsilon < y$. Thus we have

$$\mu\{S^* f > y\} \leq C \frac{\|f\|_{QA}}{y}.$$

$\square$

Finally we are going to prove the theorem of Antonov.

Theorem 12.17 (Antonov) *There exists a constant $C > 0$ such that for every measurable function $f : [-\pi, \pi] \to \mathbf{C}$ and every $y > 0$ we have*

$$\mu\{S^* f > y\} \leq C \frac{\|f\|_{\mathcal{L} \log \mathcal{L} \log \log \log \mathcal{L}}}{y}.$$

Therefore the Fourier series of every function $f \in \mathcal{L} \log \mathcal{L} \log \log \log \mathcal{L}$ is almost everywhere convergent.

Proof. Recall that we said that $f \in \mathcal{L} \log \mathcal{L} \log \log \log \mathcal{L}$ when

$$\int_X |f|(\log^+ |f|)(\log^+ \log^+ \log^+ |f|) \, d\mu < +\infty.$$

We have seen in Section 11.6 that this is a Banach space with norm

$$\|f\|_{\mathcal{L} \log \mathcal{L} \log \log \log \mathcal{L}} = \int_0^\infty f^*(t) \varphi(1/t) \, dt,$$

where $\varphi : [0, +\infty) \to [0, +\infty)$ is defined by $\varphi(t) = L_1(t) L_3(t)$. This function satisfies the conditions (1), (2), (3), and (4) of Section 11.5.

Let f be a function in $\mathcal{L} \log \mathcal{L} \log \log \log \mathcal{L}$. Let f^* be the decreasing rearrangement of f. Consider the sequence $(x_n)_{n=0}^\infty$, where $x_n = \exp(1 - e^n)$, define $a_n = f^*(x_n)$, and let $A_n = \{|f| \geq a_{n-1}\}$. We see that $A_1 = [-\pi, \pi]$ and that (A_n) is a decreasing sequence of measurable sets. We can put $f = \sum_{n=1}^\infty f_n$ where the functions f_n are defined as

$$f_n(t) = \begin{cases} f(t) - a_{n-1}\operatorname{sgn}(f(t)) & \text{if } t \in A_n \smallsetminus A_{n+1} \\ (a_n - a_{n-1})\operatorname{sgn}(f(t)) & \text{if } t \in A_{n+1}, \\ 0 & \text{if } t \notin A_n, \end{cases}$$

where $\operatorname{sgn}(a) = a/|a|$ if $a \neq 0$, and $\operatorname{sgn}(0) = 0$.

Then we have

$$e\frac{\|f_n\|_\infty}{\|f_n\|_1} \leq e\frac{\|f_n\|_\infty}{\|f_n\|_\infty \exp(1 - e^n)} = \exp(e^n).$$

Since $|f_n(t)| = |f(t)| - a_{n-1}$ if $a_n > |f(t)| \geq a_{n-1}$, $|f_n(t)| = a_n - a_{n-1}$ if $|f(t)| \geq a_n$ and $|f_n(t)| = 0$ in other case, we have

$$\{|f_n| > y\} = \begin{cases} \{|f| > y + a_{n-1}\} & \text{if } 0 < y < a_n - a_{n-1}, \\ \emptyset & \text{if } y > a_n - a_{n-1}. \end{cases}$$

Hence

$$\|f_n\|_1 = \int_0^{+\infty} \mu\{|f_n| > y\}\, dy = \int_0^{a_n - a_{n-1}} \mu\{|f| > y + a_{n-1}\}\, dy =$$

$$\int_{a_{n-1}}^{a_n} \mu\{|f| > y\}\, dy = \int_{a_{n-1}}^{a_n} x\, dy,$$

where x denotes the function $\mu_f(y)$. Therefore

$$\|f\|_{QA} \leq \sum_{n=1}^{\infty} (1 + \log n)e^n \int_{a_{n-1}}^{a_n} x\, dy.$$

Observe that when $a_{n-1} < y < a_n$ we have $x \in (x_n, x_{n-1})$ therefore

$$\exp(e^{n-1}) \leq \frac{e}{x} \leq \exp(e^n).$$

Thus for this values of y

$$e^n \leq e(\log e/x); \quad n \leq \log e(\log e/x); \quad 1 + \log n \leq \log e\big(\log e(\log e/x)\big).$$

Put $\psi(x) = (\log ex)\big(\log e(\log e(\log ex))\big)$. Then

$$\|f\|_{QA} \leq \sum_{n=1}^{\infty} e\int_{a_{n-1}}^{a_n} x\psi(1/x)\, dx = e\int_0^{+\infty} x\psi(1/x)\, dx.$$

It is not difficult to see that there exists a constant C such that for $x \geq 1$ we have $\psi(x) \leq e^{-1}CL_1(x)L_3(x) = C\varphi(x)$. Therefore we have

$$\|f\|_{QA} \leq C\int_0^{+\infty} x\varphi(x)\, dy \leq C\int_0^{+\infty}\int_0^x \big(\varphi(1/s)\, ds\big)\, dy$$

Observe that if $f^*(s) = t$ we have $s < x$ if and only if $t > y$. Therefore by Fubini's Theorem we get

$$\|f\|_{QA} \leq C \int_0^{+\infty} f^*(s)\varphi(1/s)\,ds = C\|f\|_{\mathcal{L}\log\mathcal{L}\log\log\log\mathcal{L}}.$$

□

Remark. By modifying the proof of Carleson Hunt's Theorem I get the following result.

Theorem 12.18 *There exists an absolute constant $C > 0$, such that for every $f \in \mathcal{L}^2[-\pi, \pi]$, $1 \leq p \leq 2$, and $y > 0$, we have*

$$\mu\{S^*f > y\} \leq C^p \frac{\|f\|_p^p}{y^p}\Big\{\log\Big(e\frac{\|f\|_2}{\|f\|_p}\Big)\Big\}^p,$$

In the case $p = 1$ this is a consequence of our fundamental Theorem 12.16.

Remark. We have not included all the results about the operator S^*. The most important that we have omitted are:

The result of Sjölin [40] that S^* maps $\mathcal{L}(\log\mathcal{L})^{1+\theta}$ into $\mathcal{L}(\log\mathcal{L})^{1-\theta}$, for $0 < \theta \leq 1$.

The result of Soria [43] that S^* maps $\mathcal{L}\log\mathcal{L}\log\log\mathcal{L}$ to $\mathcal{L}(\log\mathcal{L})^{-1}$.

Remark. We must complete this chapter with some facts about the negative results. In 1922 Kolmogorov, a 19 year old student of Luzin obtained the first example of a function $f \in \mathcal{L}^1(\mu)$ with almost everywhere divergent Fourier series. Four years later he gave the example of such a function with everywhere divergent Fourier series. These results were refined by Stein and Kahane. Recently Konyagin [33] has announced that given a nondecreasing function $\varphi: [0, +\infty) \rightarrow [0, +\infty)$ such that

$$\varphi(t) = o\Big(\frac{\sqrt{\log t}}{\sqrt{\log\log t}}\Big), \qquad t \rightarrow +\infty,$$

then there exists a function $f \in \mathcal{L}\varphi(\mathcal{L})$ whose Fourier series is everywhere divergent.

13. Fourier Transform on the line

13.1 Introduction

We collect here the inmediate consequences for the Fourier integral on the line. Also we give an example, that Luis Rodriguez Piazza showed to me, that proves that these results are sharp.

13.2 Fourier transform

Let $1 < p < +\infty$ and $f \in \mathcal{L}^p(\mathbf{R})$. We define two maximal operators related with the Fourier transform

$$F^* f(x) = \sup_{a>0} \left| \int_{-a}^{a} f(t) e^{-2\pi i t x} \, dt \right|,$$

$$S_{\mathbf{R}}^* f(x) = \sup_{a>0} \left| \int_{\mathbf{R}} f(t) \frac{\sin 2\pi a (x-t)}{x-t} \, dt \right|.$$

Theorem 13.1 *Let $1 < p < +\infty$. There exist constants $C_p \le C p^4/(p-1)^2$ such that for every $f \in \mathcal{L}^p(\mathbf{R})$ we have*

$$\| S_{\mathbf{R}}^* f \|_p \le C_p \| f \|_p.$$

Moreover if $1 \le p \le 2$ and p' denotes the conjugate exponent

$$\| F^* f \|_{p'} \le C_{p'} \| f \|_p.$$

Proof. Let I be any interval, and $x \in I/2$. For every real number s we have by a change of frequency

$$|C_{(s,I)} f(x) - C_{(\lfloor s \rfloor, I)} f(x)| \le C \| f \|_{(\lfloor s \rfloor, I)} \le C \| f \|_{\mathcal{L}^p(I)}.$$

It follows that

$$\sup_{s \in \mathbf{R}} |C_{(s,I)} f(x)| \le C_I^* f(x) + C \| f \|_{\mathcal{L}^p(I)}.$$

Therefore

$$\left\| \sup_{s \in \mathbf{R}} |C_{(s,I)}f(x)| \right\|_{L^p(I/2)} \leq \|C_I^* f(x)\|_{L^p(I/2)} + C \|f\|_{L^p(I)}.$$

From which we derive

$$\int_{I/2} \left(\sup_{s \in \mathbf{R}} |C_{(s,I)}f(x)| \right)^p dx \leq C_p^p \int_I |f(x)|^p \, dx.$$

If we put $I = [-2a, 2a]$, we can write this as

$$\int_{-a}^{a} \left(\sup_{s \in \mathbf{R}} \left| \text{p.v.} \int_{-2a}^{2a} f(t) \frac{e^{-\frac{2\pi i s}{4a}(x-t)}}{x-t} \, dt \right| \right)^p dx \leq C_p^p \int_{-2a}^{2a} |f(x)|^p \, dx.$$

This is the same as

$$\int_{-a}^{a} \left(C_{\mathbf{R},a}^* f(x) \right)^p dx \leq C_p^p \int_{-2a}^{2a} |f(x)|^p \, dx,$$

where

$$C_{\mathbf{R},a}^* f(x) = \sup_{s \in \mathbf{R}} \left| \text{p.v.} \int_{-2a}^{2a} f(t) \frac{e^{-2\pi i s(x-t)}}{x-t} \, dt \right|.$$

We shall consider also the following Carleson maximal operator

$$C_{\mathbf{R}}^* f(x) = \sup_{s \in \mathbf{R}} \left| \text{p.v.} \int_{\mathbf{R}} f(t) \frac{e^{2\pi i s(x-t)}}{x-t} \, dt \right|.$$

Since

$$\lim_{a \to \infty} \text{p.v.} \int_{-2a}^{2a} f(t) \frac{e^{-2\pi i s(x-t)}}{x-t} \, dt = \text{p.v.} \int_{\mathbf{R}} f(t) \frac{e^{-2\pi i s(x-t)}}{x-t} \, dt,$$

we have, for every $y > 0$

$$\frac{1}{2} \inf \left(C_{\mathbf{R}}^* f(x), y \right) \leq \liminf_{a \to +\infty} C_{\mathbf{R},a}^* f(x).$$

Therefore

$$\frac{1}{2^p} \int_{\mathbf{R}} \inf \left(C_{\mathbf{R}}^* f(x), y \right)^p dx \leq C_p^p \int_{\mathbf{R}} |f(x)|^p \, dx.$$

Hence

$$\|C_{\mathbf{R}}^* f\|_{L^p(\mathbf{R})} \leq 2C_p \|f\|_{L^p(\mathbf{R})}.$$

Now it is clear that $S_{\mathbf{R}}^* f(x) \leq C_{\mathbf{R}}^* f(x)$. Therefore

$$\|S_{\mathbf{R}}^* f\|_{L^p(\mathbf{R})} \leq 2C_p \|f\|_{L^p(\mathbf{R})}.$$

We know that the Fourier transform sends $\mathcal{L}^1(\mathbf{R})$ to $\mathcal{L}^\infty(\mathbf{R})$, and $\mathcal{L}^2(\mathbf{R})$ to $\mathcal{L}^2(\mathbf{R})$. By interpolation we can define it for $f \in \mathcal{L}^p(\mathbf{R})$, $1 \le p \le 2$. Then $\widehat{f} \in \mathcal{L}^{p'}(\mathbf{R})$ and we have the Hausdorff-Young inequality

$$\|\widehat{f}\|_{\mathcal{L}^{p'}(\mathbf{R})} \le \|f\|_{\mathcal{L}^p(\mathbf{R})}.$$

By the multiplication theorem, if $f \in \mathcal{L}^2(\mathbf{R})$

$$\int_{-a}^a f(\xi)e^{-2\pi i \xi x}\, d\xi = \int_{\mathbf{R}} \widehat{f}(t) D_a(x-t)\, dt.$$

where $D_a(t) = \sin 2\pi a t/\pi t$ is the Fourier transform of $\chi_{[-a,a]}$.
This is true also for every $f \in \mathcal{L}^p(\mathbf{R})$, $(1 \le p \le 2)$.
Hence it follows that $F^* f(x) \le C_{\mathbf{R}}^* \widehat{f}(x)$ and then

$$\|F^* f\|_{\mathcal{L}^{p'}} \le 2C_{p'}\|\widehat{f}\|_{\mathcal{L}^{p'}(\mathbf{R})} \le 2C_{p'}\|f\|_{\mathcal{L}^p(\mathbf{R})}.$$

$\square$

As a corollary we have that for every $f \in \mathcal{L}^p(\mathbf{R})$, with $1 \le p \le 2$ we have

$$\widehat{f}(x) = \lim_{a \to +\infty} \int_{-a}^a f(t) e^{-2\pi i t x}\, dt, \qquad \text{a. e. on } \mathbf{R}.$$

The inequality $\|F^* f\|_{p'} \le C_{p'}\|f\|_p$ of the preceding theorem can not be extended to values of $p > 2$ as we see in the following example.

Example 13.2 *There is a function $f\colon \mathbf{R} \to \mathbf{R}$ such that $f \in \mathcal{L}^p(\mathbf{R})$ for every $p > 2$ and such that*

$$\sup_{a>0}\left|\int_{-a}^a f(t) e^{2\pi i t x}\, dt\right| = +\infty, \qquad a.e. \text{ on } \mathbf{R}.$$

Proof. Let f be the function

$$f(t) = \sum_{n=1}^\infty \frac{1}{\sqrt{n}} \chi_{[2^n, 2^n+1]}(t).$$

It is obvious that $f \in \mathcal{L}^p(\mathbf{R})$ for every $p > 2$.
On the other hand the Fourier transform of $\chi_{[2^n, 2^n+1]}(t)$ is equal to

$$\frac{\sin \pi x}{\pi x} e^{-\pi i x} e^{-2\pi i 2^n x}.$$

It follows easily that

$$F^* f(x) = \sup_{a>0} \left| \int_{-a}^{a} f(t) e^{2\pi i t x} \, dt \right| \geq \left| \frac{\sin \pi x}{\pi x} \right| \cdot \sup_{N \geq 1} \left| \sum_{n=1}^{N} \frac{1}{\sqrt{n}} e^{-2\pi i 2^n x} \right|.$$

Consequently, to prove our assertion it is enough to show that the last supremum is infinite a.e.

The set of powers of 2 is a Λ_4 set, that is for every trigonometric polynomial of the form $P(x) = \sum_{n=1}^{N} a_n e^{-2\pi i 2^n x}$ we have $\|P\|_4 \leq 2^{1/4} \|P\|_2$ in fact

$$\|P\|_4^4 = \int_0^1 \left(\sum_{n=1}^{N} a_n e^{-2\pi i 2^n x} \right)^2 \left(\sum_{n=1}^{N} \overline{a_n} e^{2\pi i 2^n x} \right)^2 dx$$

and since $2^k + 2^l = 2^n + 2^m$ only if $\{k, l\} = \{n, m\}$ we obtain

$$\|P\|_4^4 = 4 \sum_{n \neq m} |a_n|^2 |a_m|^2 + \sum_{n} |a_n|^4 \leq 2 \left(\sum_{n} |a_n|^2 \right)^2 = 2\|f\|_2^4.$$

As a consequence of the equivalence of the two norms, there is some absolute constant $\delta > 0$ such that for every trigonometric polynomial in powers of 2, we have

$$\mathfrak{m}\{x \in [0,1] : |P(x)| \geq \|P\|_2/2\} > \delta.$$

To see this, fix $\theta \in (0,1)$, and let $A = \{x \in [0,1] : |P(x)| \geq \theta \|P\|_2\}$. We have

$$(1 - \theta^2) \|P\|_2^2 \leq \int_A |P(x)|^2 \, d\mu \leq \sqrt{\mathfrak{m}(A)} \left(\int_A |P|^4 \, dm \right)^{1/2}$$
$$\leq \sqrt{\mathfrak{m}(A)} \|P\|_4^2 \leq C \sqrt{\mathfrak{m}(A)} \|P\|_2^2.$$

Therefore $\mathfrak{m}(A) \geq (1 - \theta^2)^2 / C^2$.

Let B be the set of $x \in \mathbf{R}$ where

$$\sup_{N \geq 1} \left| \sum_{n=1}^{N} \frac{1}{\sqrt{n}} e^{-2\pi i 2^n x} \right| = +\infty.$$

This set B is periodic with period 2^n for every $n \in \mathbf{N}$, because it coincides with the set where

$$\sup_{N \geq n_0} \left| \sum_{n=n_0}^{N} \frac{1}{\sqrt{n}} e^{-2\pi i 2^n x} \right| = +\infty.$$

It follows that $\mathfrak{m}(B \cap [0,1])$ must be 0 or 1. The set B contains the limit superior of the sets

$$B_N = \left\{ x \in [0,1] : \left| \sum_{n=1}^{N} \frac{1}{\sqrt{n}} e^{-2\pi i 2^n x} \right| \geq \frac{1}{2} \sqrt{\log N} \right\}.$$

Hence $\mathfrak{m}(B \cap [0,1]) \geq \delta > 0$. It follows that $\mathbf{R} \setminus B$ is of Lebesgue measure zero. $\square$

References

[1] N. Yu. Antonov: Convergence of Fourier series, East J. Approx., **2**, 187–196 (1966)

[2] N. A. Bary: Treatise on Trigonometric Series, vols. 1 and 2, Pergamon Press Inc, New York, 1964

[3] C. Bennett, R. Sharpley: Interpolation of Operators, Academic Press, Boston 1988

[4] J. Bergh, J. Löfström: Interpolation spaces, Springer, Berlin 1976

[5] A. S. Besikovitch: Sur la nature des fonctions à carré sommable et des ensembles measurables, Fund. Math., **4**, 172–195 (1923)

[6] A. S. Besikovitch: On a general metric property of summable functions, J. London Math. Soc., **1**, 120–128 (1926)

[7] P. Billard: Sur la convergence presque partout des séries de Fourier-Walsh des fonctions de l'espace $L^2(0, 1)$, Studia Math., **28**, 363–388 (1967)

[8] A. P. Calderón, A. Zygmund: On the existence of certain singular integrals, Acta Math., **88**, 85–139 (1952)

[9] L. Carleson: On convergence and growth of partial sums of Fourier series, Acta Math., **116**, 135–157 (1966)

[10] Y. M. Chen: An almost everywhere divergent Fourier series of the class $L(\log^+ \log^+ L)^{1-\epsilon}$, J. London Math. Soc., **44**, 643–654 (1969)

[11] J. Duoandikoetxea: Fourier Analysis, American Math. Soc., Providence, RI, 2001

[12] R. E. Edwards: Fourier Series a Modern Introduction, vols. 1 and 2, Holt Rinehart and Winston Inc, New York 1967

[13] P. Fatou: Séries trigonométriques et séries de Taylor, cta Math., **30**, 335–400 (1906)

[14] C. Fefferman: Pointwise convergence of Fourier series, Ann. of Math., **98**, 551–571 (1973)

[15] C. Fefferman: Erratum Pointwise convergence of Fourier series, Ann. of Math., **146**, 239 (1997).

[16] G. Gallavotti, A. Porzio: The almost everywhere convergence of Fourier series, Quaderno del Consiglio Nazionale dell Ricerche, Grupo Nazionale di Fisica Matematica, Roma 1987

[17] A. Garsia: Topics in almost everywhere Convergence, Markham Publ., Chicago 1970

[18] M. de Guzmán: Differentiation of Integrals in $\mathbf{R}^n$, Springer, Lect. Notes in Math. 481, Berlin 1975.

[19] M. de Guzmán: Real Variable Methods in Fourier Analysis, North Holland, Amsterdam, 1981

[20] G. H. Hardy: On the summability of Fourier's series, Proc. London Math. Soc., **12**, 365–372 (1913)

[21] G. H. Hardy, J. E. Littlewood: A maximal theorem with function-theoretic applications, Acta Math., **54**, 81–116 (1930)

[22] R. A. Hunt: On $L(p,q)$ spaces, L'Enseignement Mathematique, **12**, 249–276 (1966).

[23] R. A. Hunt: On the convergence of Fourier series. Orthogonal Expansions and their continuous analogues, Proc. Conf. Edwardsville (1967), Ill, Southern Illinois Univ. Press, 235–255 (1966).

[24] R. A. Hunt: Comments on Luzin's conjecture and Carleson's proof for L^2 Fourier series, Linear Operators and approximation, II (Proc. Conf., Oberwolfach Math. Res. Inst., Oberwolfach 1974), Internat. Ser. Numer. Math., Birkhauser, Basel 235–245 (1974)

[25] R. A. Hunt, M. H. Taibleson: Almost everywhere convergence of Fourier series on the ring of integers of a local field, Siam J. Math. Anal., **2**, 607–625 (1971)

[26] O. G. Jørsboe, L. Mejlbro: The Carleson-Hunt theorem on Fourier Series, Springer, Lect. Notes in Math. 911, Berlin 1982.

[27] J. P. Kahane: Sur la divergence presque sûre partout de certaines séries de Fourier aléatoires, Ann. Univ. Sci. Budapest, Eötvös Sect. Math., **3–4**, 101–108 (1960/61)

[28] N. J. Kalton: Convexity, type and the three space problem, Studia Math., **69**, 247–287 (1981)

[29] Y. Katznelson: An Introduction to Harmonic Analysis, Dover, New York 1976

[30] A. Kolmogorov: Une série de Fourier-Lebesgue divergente presque partout, Fund. Math., **4**, 324–328 (1923)

[31] A. Kolmogorov: Une série de Fourier-Lebesgue divergente partout, C. R. Acad. Sci. Paris, **183**, 1327–1329 (1926)

[32] S. V. Konyagin: On divergence of trigonometric Fourier series everywhere, C. R. Acad. Sci. Paris, **329**, 693–697 (1999)

[33] S. V. Konyagin: On the almost everywhere divergence of Fourier series, Matematicheskii Sbornik, **191**, 103–126 (2000). Translation in Sb. Math. **191**, 361–370 (2000)

[34] T. W. Körner: Everywhere divergent Fourier series, Colloq. Math., **45**, 103–118 (1981)

[35] M. Lacey, C. Thiele: A proof of boundedness of the Carleson operator, Math. Res. Lett., **7**, 361–370 (2000)

[36] N. Luzin: Sur la convergence des séries trigonométriques de Fourier, C. R. Acad. Sci. Paris, **156**, 1655–1658 (1913)

[37] A Mátté: The convergence of Fourier series of square integrable functions, Matematikai Lapok, **18**, 195–242 (1967)

[38] C. J. Mozzochi: On the pointwise convergence of Fourier Series, Springer, Lect. Notes in Math. 199, Berlin 1971.

[39] P. Sjölin: An inequality of Paley and convergence a.e. of Walsh-Fourier series, Arkiv Math., **7**, 551–570 (1969)

[40] P. Sjölin: Two theorems in Fourier integrals and Fourier series, in Approximations and Function spaces, Banach Center Publ., **22**, 413–426 PWN, Warsaw 1989

[41] F. Soria: Note on differentiation of integrals and the halo conjecture, Studia Math., **81**, 29–36 (1985)

[42] F. Soria: On an extrapolation theorem of Carleson-Sjölin with applications to a.e. convergence of Fourier series, Studia Math., **94**, 235–244 (1989)

[43] F. Soria: Integrability properties of the maximal operator on partial sums of Fourier series, Rend. Circ. Mat. Palermo, **38**, 371–376 (1989)

[44] E. M. Stein: On limits of sequences of operators, Ann. Math., **74**, 140–170 (1961)

[45] E. M. Stein: Singular integrals and differentiability properties of functions, Princeton University Press, Princeton 1970

[46] E. M. Stein, G. Weiss: Introduction to Fourier analysis on euclidean spaces, Princeton University Press, Princeton 1971

[47] E. C. Titchmarsh: Reciprocal Formulae involving series and integrals, Math. Z., **25**, 321–347 (1926)

[48] A. Torchinski: Real-Variable Methods in Harmonic Analysis, Academic Press, New York 1986

[49] S. Yano: Notes on Fourier Analysis (XXIX): An extrapolation theorem, J. Math. Soc. Japan, **3**, 296–305 (1951)

[50] A. Zygmund: Trigonometric Series, Cambridge University Press, New York 1959

Comments

The theorem about the pointwise almost everywhere convergence of Poisson integrals of functions in $\mathcal{L}^1$ (theorem 2.10) is contained in [13].

In [20] is proved theorem 2.5. Here Hardy also asks whether the estimate $o(\log n)$ is best possible.

Luzin [36] asked for a *real* proof (instead of *complex*) of the existence of Hilbert transform for $f \in \mathcal{L}^2$, and conjectured Carleson's Theorem. The first author to give a *real* proof of the existence of the Hilbert transform was Besikovitch [5]. Then Titchmarsh [47] gave one for $f \in \mathcal{L}^p$, $p > 1$. Finally Besikovitch [6] obtained the proof in the case $p = 1$. A generalization of all these results is obtained in the seminal paper of Calderón and Zygmund [8].

An important tool for the Carleson-Hunt Theorem is the maximal function of Hardy and Littlewood [21].

Some important consequences about the almost everywhere convergence of Fourier series were obtained by Stein [44].

The first example of a function $f \in \mathcal{L}^1$ with a. e. divergent Fourier series was given by Kolmogorov [30]. Four years later he obtained an everywhere divergent Fourier series [31]. We can see the construction and some consequences in Körner [34]. Related results are contained in the papers by Kahane [27] and Chen [10]. The latest results about divergent Fourier series have been announced by Konyagin [33].

The proof of Carleson's Theorem appeared in Acta. Math. [9]. In the same year was published a paper of Hunt [22] where he reviewed the interpolation properties of the $\mathcal{L}^{p,q}$ spaces. This paper contains parts of the Ph. D. Thesis of Hunt under the direction of G. Weiss.

In 1968 Hunt [23] obtained our theorems 12.1, 12.5, 12.6 and 12.8. But the constant he gave in theorem 12.8 is $Cp^5/(p-1)^3$. Many interesting historical notes are given in another paper by Hunt [24].

The first modification of the Carleson Theorem was given by Billard [7] to the series of Walsh. The extension to the case of Walsh-Fourier series of the Theorem of Hunt is given by Sjölin [39]. In this paper it is proved also that the Fourier series of a function in $\mathcal{L} \log \mathcal{L} \log \log \mathcal{L}$ is a.e. convergent.

In Hunt and Taibleson [25] the Carleson-Hunt theorem is extended to the $\mathcal{L}^p$ space of the integers of a local field endowed with the Haar measure. The paper [25] contains also a simplified proof of the result of Sjölin, which is done by using special functions (those that can be written as $f = g\,\chi_A$, with $1/2 \leq g \leq 1$).

The first theorem of extrapolation was given by Yano [49]. He proves that if an operator is bounded on $\mathcal{L}^p$, $(1 < p < p_0)$ with norm $O((p-1)^r)$, then it maps $\mathcal{L}(\log \mathcal{L})^r$ to $\mathcal{L}^1$. This was refined and applied to the operator S^* giving the results of Sjölin that we have mentioned.

Soria [41] and [42] defined a quasi-Banach function space B_1, and proved that every function in this space has an a.e. convergent Fourier series. This space is similar to our quasi-Banach space Q. We have applied a result of Kalton [28] to define the quasi-Banach spaces Q and QA. In another direction Sjölin [39] proved that S^* maps $\mathcal{L}(\log \mathcal{L})^{1+\theta}$ into $\mathcal{L}(\log \mathcal{L})^{1-\theta}$ for $0 < \theta \leq 1$. And finally Soria [43] gave the final form to the Sjölin result. He proved essentially $\|f\|_Q \leq C\|f\|_{\mathcal{L}\log\mathcal{L}\log\log\mathcal{L}}$.

The result of Antonov is given in his paper [1].

Other proofs of Carleson's Theorem have been given. First was one by Fefferman [14], (see Gallavotti and Porzio [16] for an exposition). Recently there has appeared another short proof by Lacey and Thiele [35].

We have consulted many books: Zygmund [50], Bary [2], Edwards [12], Katznelson [29], Stein [45], Stein-Weiss [46], and Duoandikoetxea [11], Guzmán [18] and [19], Torchinski [48], Garsia [17], Bergh y Löfström [4], and Bennett and Sharpley [3].

We know of two previous treatments of the Carleson Hunt theorem, one by Mozzochi [38] and the other by Jørsboe and L. Mejlbro [26]. Also we have noticed that there is one in Hungarian by Mátté [37].

Subject Index

Recent Reprints and New Editions

4. Lecture Notes are printed by photo-offset from the master-copy delivered in camera-ready form by the authors. Springer-Verlag provides technical instructions for the preparation of manuscripts. Macro packages in $\TeX$, $\LaTeX$2e, $\LaTeX$2.09 are available from Springer's web-pages at

http://www.springer.de/math/authors/b-tex.html.

Careful preparation of the manuscripts will help keep production time short and ensure satisfactory appearance of the finished book.

The actual production of a Lecture Notes volume takes approximately 12 weeks.

5. Authors receive a total of 50 free copies of their volume, but no royalties. They are entitled to a discount of 33.3% on the price of Springer books purchase for their personal use, if ordering directly from Springer-Verlag.

Commitment to publish is made by letter of intent rather than by signing a formal contract. Springer-Verlag secures the copyright for each volume. Authors are free to reuse material contained in their LNM volumes in later publications: A brief written (or e-mail) request for formal permission is sufficient.

Addresses:

Professor Jean-Michel Morel
CMLA, École Normale Supérieure de Cachan
61 Avenue du Président Wilson
94235 Cachan Cedex France
e-mail: Jean-Michel.Morel@cmla.ens-cachan.fr

Professor Bernard Teissier
Institut de Mathématiques de Jussieu
Equipe "Géométrie et Dynamique"
175 rue du Chevaleret
75013 PARIS
e-mail: Teissier@ens.fr

Professor F. Takens, Mathematisch Instituut
Rijksuniversiteit Groningen, Postbus 800
9700 AV Groningen, The Netherlands
e-mail: F.Takens@math.rug.nl

Springer-Verlag, Mathematics Editorial, Tiergartenstr. 17
D-69121 Heidelberg, Germany
Tel.: +49 (6221) 487-701
Fax: +49 (6221) 487-355
e-mail: lnm@Springer.de